# DeWALT®

# CONSTRUCTION SAFETY/OSHA

## PROFESSIONAL REFERENCE

Paul R

Created exclusively
for DeWALT by:

# PAL
publications®

www.palpublications.com
1-800-246-2175

## OTHER TITLES AVAILABLE

<u>**Trade Reference Series**</u>

- Blueprint Reading
- Construction
- Construction Estimating
- Datacom
- Electric Motor
- Electrical
- Electrical Estimating
- HVAC/R – Master Edition
- Lighting & Maintenance
- Plumbing
- Residential Remodeling & Repair
- Security, Sound & Video
- Spanish/English Construction Dictionary – Illustrated
- Wiring Diagrams

<u>**Exam and Certification Series**</u>

- Building Contractor's Licensing Exam Guide
- Electrical Licensing Exam Guide
- HVAC Technician Certification Exam Guide
- Plumbing Licensing Exam Guide

For a complete list of The DEWALT Professional Trade Reference Series visit **www.dewalt.com/guides**.

**This Book Belongs To:**

Name:_____

Company: _____

Title: _____

Department: _____

Company Address: _____

_____

_____

Company Phone: _____

Home Phone: _____

Pal Publications, Inc.
800 Heritage Drive, Suite 810
Pottstown, PA 19464-3810

ISBN 0-9777183-3-6
10  09  08  07  06      5  4  3  2  1
Printed in the United States of America

# Preface

Safety has always been a serious issue in the construction business as well as the industry in general. Safety is both a moral responsibility and an economic one, with injuries simply being bad business. There has always been a legal aspect to safety as well (reckless endangerment having been a recognized crime for centuries), but with the introduction of OSHA in the United States, it has become a very obvious legal issue.

This book covers all the major aspects of safety on construction sites. The most crucial safety threats such as scaffolds, ladders, electrical, and excavation are addressed in their own, individual chapters.

Chapter Four – Worker Safety Instructions and Chapter Nine – Company Safety Plan are unique in that they can be copied, enlarged, and used in daily operations. I think that this will prove to be highly beneficial, especially to smaller contractors who do not have the time and money to put together programs from scratch.

I have also included a glossary with all the major construction and OSHA terms, plus necessary conversion factors.

Naturally, there may be some aspects of job site safety and/or OSHA requirements that I have not covered in sufficient depth for some readers. I will update this book on a continual basis and will attempt to include material suggested by readers and keep pace with developments in the trades.

Best wishes,
Paul Rosenberg

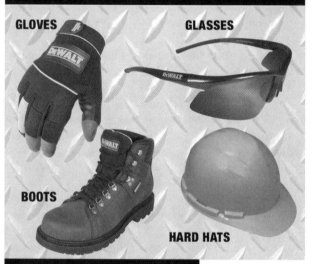

# CONTENTS

## CHAPTER 1 – *Construction Sites* . . . . . 1-1

# CHAPTER 2 – *Scaffolding and Ladders*..................... 2-1

# CHAPTER 5 – *Excavation* . . . . . . . . 5-1

# CHAPTER 6 – *Electrical Safety* . . . . . . . 6-1

# CHAPTER 9 – *Company Safety Plan* . . 9-1

## CHAPTER 10 – *Glossary and Abbreviations* . . . . . . . . . . . . . . . . . . . . 10-1

# CHAPTER 1
## *Construction Sites*

### FALL PROTECTION

Each year, falls consistently account for the greatest number of fatalities in the construction industry. A number of factors are often involved in falls, including unstable working surfaces, misuse or failure to use fall protection equipment and human error. Guardrails, fall arrest systems, safety nets, covers and restraint systems can prevent many deaths and injuries from falls.

- Use aerial lifts or elevated platforms to provide safer elevated working surfaces;
- Erect guardrail systems with toeboards and warning lines or install control line systems to protect workers near the edges of floors and roofs;
- Cover floor holes; and/or
- Use safety net systems or personal fall arrest systems (body harnesses).

### HEAD PROTECTION

Serious injuries can result from blows to the head. Be sure that workers wear hard hats where there is a potential for objects falling from above, bumps to their heads from fixed objects or accidental head contact with electrical hazards.

## SCAFFOLDING

When scaffolds are not erected or used properly, fall hazards can occur. About 2.3 million construction workers frequently work on scaffolds.

- Scaffolds must be sound, rigid and sufficient to carry their own weight plus four times the maximum intended load without settling or displacement. They must be erected on solid footing.

- Unstable objects, such as barrels, boxes, loose bricks or concrete blocks must not be used to support scaffolds or planks.

- Scaffolds must not be erected, moved, dismantled or altered except under the supervision of a competent person.

- Scaffolds must be equipped with guardrails, midrails and toeboards.

- Scaffold accessories such as braces, brackets, trusses, screw legs or ladders that are damaged or weakened from any cause must be immediately repaired or replaced.

- Scaffold platforms must be tightly planked with scaffold plank–grade material or the equivalent.

- A competent person must inspect the scaffolding and, at designated intervals, reinspect it.

- Rigging on suspension scaffolds must be inspected by a competent person before each shift and after any occurrence that could affect structural integrity to ensure that all connections are tight and that no damage to the rigging has occurred since its last use.

## SCAFFOLDING *(cont.)*

- Synthetic and natural rope used in suspension scaffolding must be protected from heat-producing sources.
- Employees must be instructed about the hazards of using diagonal braces as fall protection.
- Scaffolds must be accessible by ladders and stairwells.
- Scaffolds must be at least 10' from electric power lines at all times.

## LADDERS

Ladders and stairways are another source of injuries and fatalities among construction workers. OSHA estimates that there are 24,882 injuries and as many as 36 fatalities per year nationwide due to falls on stairways and ladders used in construction. Nearly half of these injuries were serious enough to require time off the job.

- Use the correct ladder for the task.
- Have a competent person visually inspect a ladder before use for any defects such as:
  - Structural damage, split/bent side rails, broken or missing rungs/steps/cleats and missing or damaged safety devices;
  - Grease, dirt or other contaminants that could cause slips or falls; or
  - Paint or stickers (except warning labels) that could hide possible defects.

## LADDERS *(cont.)*

- Make sure that ladders are long enough to safely reach the work area.
- Mark or tag **"Do Not Use"** damaged or defective ladders for repair or replacement, or destroy them immediately.
- Never load ladders beyond the maximum intended load or beyond the manufacturer's rated capacity.
- Be sure the load rating can support the weight of the user, including materials and tools.
- Avoid using ladders with metallic components near electrical work and overhead power lines.

## STAIRWAYS

Slips, trips and falls on stairways are a major source of injuries and fatalities among construction workers.

- Stairway treads and walkways must be free of dangerous objects, debris and materials.
- Slippery conditions on stairways and walkways must be corrected immediately.
- Make sure that treads cover the entire step and landing.
- Stairways having four or more risers or rising more than 30" must have at least one handrail.

## HAZARD COMMUNICATION

Failure to recognize the hazards associated with chemicals can cause chemical burns, respiratory problems, fires and explosions.

- Maintain a Material Safety Data Sheet (MSDS) for each chemical in the facility.
- Make this information accessible to employees at all times in a language or format that is clearly understood by all affected personnel.
- Train employees on how to read and use the MSDS.
- Follow the manufacturer's MSDS instructions for handling hazardous chemicals.
- Train employees about the risks of each hazardous chemical being used.
- Provide spill clean-up kits in areas where chemicals are stored.
- Have a written spill control plan.
- Train employees to clean up spills, protect themselves and properly dispose of used materials.
- Provide proper personal protective equipment and enforce its use.
- Store chemicals safely and securely.

# GROUND FAULT INTERRUPTERS

It is the employer's responsibility to provide either:

1. Ground-fault circuit interrupters on construction sites for receptacle outlets in use and not part of the permanent wiring of the building or structure; or

2. A scheduled and recorded assured equipment grounding conductor program on construction sites, covering all cord sets, receptacles that are not part of the permanent wiring of the building or tructure and equipment connected by cord and plug that are available for use or used by employees.

In practice, option #1 above is the general method of compliance.

# EQUIPMENT GROUNDING CONDUCTOR PROGRAM

**The employer must provide the employees the following to ensure a safe program:**

- A written description of the program which includes other languages (if necessary).
- A competent person to implement the program who is entirely familiar with equipment grounding.
- An overall inspection and testing of all equipment grounding conductors.
- Written records of the test results.

## EQUIPMENT GROUNDING CONDUCTOR PROGRAM (cont.)

### Inspections

- **Frequency of inspections:**
  - Before each day's use.

- **Visual inspection of the following equipment is required:**
  - Cord sets.
  - Cap, plug and receptacle of cord sets.
  - Equipment connected by cord and plug.

- **Exceptions:**
  - Receptacles and cord sets that are fixed and not exposed to damage.

### Tests

- **Frequency of tests:**
  - Before first use.
  - After repair and before placing back in service.
  - Before use after suspected damage.
  - Every 3 months, cord sets and receptacles exposed to damage must be tested at regular intervals not to exceed 6 months.

- **Conduct tests for:**
  - Continuity of grounding equipment conductor.
  - Proper terminal connection of grounding conductor equipment.

## CONDUCTOR COLOR CODE

**Grounded Conductor**
- White
- Gray
- Three continuous white stripes

**Ungrounded Conductor**
- Any color other than white, gray or green

**Equipment Grounding Conductor**
- Green with one or more yellow stripes
- Bare

## POWER WIRING COLOR CODE

| 120/240 Volt | | 277/480 Volt | |
|---|---|---|---|
| Black | Phase 1 | Brown | Phase 1 |
| Red | Phase 2 | Orange | Phase 2 |
| Blue | Phase 3 | Yellow | Phase 3 |
| White or with three white stripes | Neutral | Gray or with three white stripes | Neutral |
| Green | Ground | Green with yellow stripe | Ground |

## POWER-TRANSFORMER COLOR CODE

| Wire Color | Transformer Circuit Type |
|---|---|
| Black | If a transformer does not have a tapped primary, both leads are black. |
| Black | If a transformer does have a tapped primary, the black is the common lead. |
| Black and Yellow | Tap for a tapped primary. |
| Black and Red | End for a tapped primary. |

## SIZE OF EXTENSION CORDS FOR PORTABLE TOOLS

| Cord Length (feet) | Full-Load Rating of the Tool in Amperes at 115 Volts | | | | | |
| --- | --- | --- | --- | --- | --- | --- |
| | 0 to 2.0 | 2.1 to 3.4 | 3.5 to 5.0 | 5.1 to 7.0 | 7.1 to 12.0 | 12.1 to 16.0 |
| | Wire Size (AWG) | | | | | |
| 25 | 18 | 18 | 18 | 16 | 14 | 14 |
| 50 | 18 | 18 | 18 | 16 | 14 | 12 |
| 75 | 18 | 18 | 16 | 14 | 12 | 10 |
| 100 | 18 | 16 | 14 | 12 | 10 | 8 |
| 200 | 16 | 14 | 12 | 10 | 8 | 6 |
| 300 | 14 | 12 | 10 | 8 | 6 | 4 |
| 400 | 12 | 10 | 8 | 6 | 4 | 4 |
| 500 | 12 | 10 | 8 | 6 | 4 | 2 |
| 600 | 10 | 8 | 6 | 4 | 2 | 2 |
| 800 | 10 | 8 | 6 | 4 | 2 | 1 |
| 1000 | 8 | 6 | 4 | 2 | 1 | 0 |

1-9

## TRENCHING

Trench collapses cause dozens of fatalities and hundreds of injuries each year. Note the following Safety Guidelines for Trenching:

- Never enter an unprotected trench.

- Always use a protective system for trenches 5' deep or greater.

- Employ a registered professional engineer to design a protective system for trenches 20' deep or greater.

- Use the following protective systems:

  A. Sloping to protect workers, by cutting back the trench wall at an angle inclined away from the excavation not steeper than a height/depth ratio of 1½:1, according to the sloping requirements for the type of soil.

  B. Shoring to protect workers by installing supports to prevent soil movement for trenches that do not exceed 20' in depth.

  C. Shielding to protect workers by using trench boxes or other types of supports to prevent soil cave-ins.

## TRENCHING *(cont.)*

- Always provide a way to exit a trench—such as a ladder, stairway or ramp—that is no more than 25' of lateral travel for employees in the trench.

- Keep spoils at least 2' back from edge of trench.

- Make sure that trenches are inspected by a competent person prior to entry and after any hazard-increasing event such as a rainstorm, vibrations or excessive surcharge loads.

## ALLOWABLE SLOPES

**Sloping.** Maximum allowable slopes for excavations less than 20' (6.09 m) based on soil type and angle to the horizontal are as follows:

| Soil Type | Height/Depth Ratio | Slope Angle |
|---|---|---|
| Stable Rock (granite or sandstone) | Vertical | 90° |
| Type A (clay) | ¾:1 | 53° |
| Type B (gravel, silt) | 1:1 | 45° |
| Type C (sand) | 1½:1 | 34° |
| Type A (short-term) (For a maximum excavation depth of 12') | ½:1 | 63° |

## CRANES

Significant and serious injuries may occur if cranes are not inspected before use and if they are not used properly. Often these injuries occur when a worker is struck by an overhead load or caught within the crane's swing radius. Many crane fatalities occur when the boom of a crane or its load line contact an overhead power line.

### Solutions:

- Check all crane controls to ensure proper operation before use.
- Inspect wire rope, chains and hook for any damage.
- Know the weight of the load that the crane is to lift.
- Ensure that the load does not exceed the crane's rated capacity.
- Raise the load a few inches to verify balance and the effectiveness of the brake system.
- Check all rigging prior to use; do not wrap hoist ropes or chains around the load.
- Fully extend outriggers.
- Do not move a load over workers.
- Barricade accessible areas within the crane's swing radius.
- Watch for overhead electrical distribution and transmission lines and maintain a safe working clearance of at least 10' from energized electrical lines.

# STANDARD HAND SIGNALS FOR CONTROLLING CRANE OPERATIONS

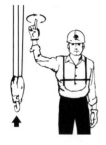

**HOIST.** Forearm vertical, forefinger pointing up, move hand in small horizontal circles.

**LOWER.** Arm extended downward, forefinger pointing down, move hand in small horizontal circles.

**USE MAIN HOIST.** Tap fist on head; then use regular signals.

**USE WHIPLINE.** Tap elbow with one hand; then use regular signals.

**RAISE BOOM.** Arm extended, fingers closed, thumb pointing upward.

**LOWER BOOM.** Arm extended, fingers closed, thumb pointing down.

**MOVE SLOWLY.** One hand gives motion signal, other hand motionless in front of hand giving the motion signal.

**RAISE THE BOOM AND LOWER THE LOAD.** Arm extended, thumb pointing up, flex fingers in and out.

**LOWER THE BOOM AND RAISE THE LOAD.** Arm extended, thumb pointing down, flex fingers in and out.

# STANDARD HAND SIGNALS FOR
# CONTROLLING CRANE OPERATIONS *(cont.)*

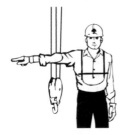

**SWING.** Arm extended, point with finger in direction of swing.

**STOP.** Arm extended, palm down, hold.

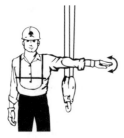

**EMERGENCY STOP.** Arm extended, palm down, move hand rapidly right and left.

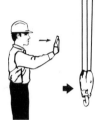

**TRAVEL.** Arm extended forward, hand open and slightly raised, pushing motion in direction of travel.

**EXTEND BOOM.** Both fists in front of body with thumbs pointing outward.

**RETRACT BOOM.** Both fists in front of body with thumbs pointing toward each other.

# MANILA ROPE SLINGS — STRAIGHT LEG

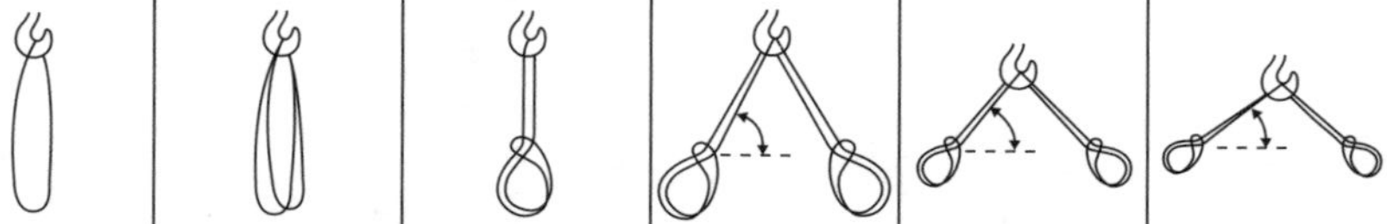

| Rope Diameter (inches) | Straight 2 Leg | Straight 4 Leg | Choker 2 Leg | 60° Choker 4 Leg | 45° Choker 4 Leg | 30° Choker 4 Leg |
|---|---|---|---|---|---|---|
| | | | TONS CAPACITY | | | |
| ½ | ½ | 1 | ⅓ | ⅔ | ½ | ⅓ |
| ¾ | ¾ | 1½ | ¾ | 1¼ | 1 | ¾ |
| 1 | 1½ | 3 | 1¼ | 2 | 1½ | 1¼ |
| 1½ | 3 | 6 | 2 | 4 | 3 | 2 |
| 2 | 5 | 10 | 4 | 7 | 6 | 4 |
| 2½ | 7 | 15 | 6 | 10 | 8 | 6 |
| 3 | 10 | 20 | 8 | 14 | 12 | 8 |
| 3½ | 14 | 29 | 11 | 20 | 16 | 11 |
| 4 | 17 | 34 | 13 | 23 | 19 | 13 |

# MANILA ROPE SLINGS – BASKET

| Rope Diameter (inches) | 60° Basket 4 Leg | 45° Basket 4 Leg | 30° Basket 4 Leg | 60° Basket 6 Leg | 45° Basket 6 Leg | 30° Basket 6 Leg |
|---|---|---|---|---|---|---|
| | | | TONS CAPACITY | | | |
| 1/2 | 3/4 | 2/3 | 1/2 | 1 | 3/4 | 2/3 |
| 3/4 | 1 1/2 | 1 | 3/4 | 2 1/4 | 2 | 1 |
| 1 | 2 1/2 | 2 | 1 1/2 | 4 | 3 | 2 |
| 1 1/2 | 5 | 4 | 3 | 8 | 6 | 4 |
| 2 | 9 | 7 | 5 | 13 | 11 | 7 |
| 2 1/2 | 13 | 11 | 7 | 19 | 16 | 11 |
| 3 | 18 | 15 | 10 | 27 | 22 | 15 |
| 3 1/2 | 25 | 21 | 15 | 38 | 31 | 22 |
| 4 | 30 | 24 | 17 | 44 | 36 | 25 |

1-19

# ROPE CHARACTERISTICS

| Rope Diameter (in.) | Safe Load Ratio | Nylon | | Polypropylene | | Manila | |
|---|---|---|---|---|---|---|---|
| | | Break Lbs. | Lbs./ 100 Feet | Break Lbs. | Lbs./ 100 Feet | Break Lbs. | Lbs./ 100 Feet |
| $3/16$ | 10:1 | 1000 | 1.0 | 800 | 0.7 | 406 | 1.5 |
| $1/4$ | 10:1 | 1650 | 1.5 | 1250 | 1.2 | 540 | 2.0 |
| $5/16$ | 10:1 | 2550 | 2.5 | 1900 | 1.8 | 900 | 2.9 |
| $3/8$ | 10:1 | 3700 | 3.5 | 2700 | 2.8 | 1220 | 4.1 |
| $7/16$ | 10:1 | 5000 | 5.0 | 3500 | 3.8 | 1580 | 5.3 |
| $1/2$ | 9:1 | 6400 | 6.5 | 4200 | 4.7 | 2380 | 7.5 |
| $9/16$ | 8:1 | 8000 | 8.3 | 5100 | 6.1 | 3100 | 10.4 |
| $5/8$ | 8:1 | 10400 | 10.5 | 6200 | 7.5 | 3960 | 13.3 |
| $3/4$ | 7:1 | 14200 | 14.5 | 8500 | 10.7 | 4860 | 16.7 |
| $13/16$ | 7:1 | 17000 | 17.0 | 9900 | 12.7 | 5850 | 19.5 |
| $7/8$ | 7:1 | 20000 | 20.0 | 11500 | 15.0 | 6950 | 22.4 |
| 1 | 7:1 | 25000 | 26.4 | 14000 | 18.0 | 8100 | 27.0 |
| $1 1/16$ | 7:1 | 28800 | 29.0 | 16000 | 20.4 | 9450 | 31.2 |
| $1 1/8$ | 7:1 | 33000 | 34.0 | 18300 | 23.8 | 10800 | 36.0 |
| $1 1/4$ | 7:1 | 37500 | 40.0 | 21000 | 27.0 | 12200 | 41.6 |
| $1 5/16$ | 7:1 | 43000 | 45.0 | 23500 | 30.4 | 13500 | 47.8 |
| $1 1/2$ | 7:1 | 53000 | 55.0 | 29700 | 38.4 | 16700 | 60.0 |
| $1 5/8$ | 7:1 | 65000 | 66.5 | 36000 | 47.6 | 20200 | 74.5 |
| $1 3/4$ | 7:1 | 78000 | 83.0 | 43000 | 59.0 | 23800 | 89.5 |
| 2 | 7:1 | 92000 | 95.0 | 52000 | 69.0 | 28000 | 108 |
| $2 1/8$ | 7:1 | 106000 | 109 | 61000 | 80.0 | | |
| $2 1/4$ | 6:1 | 125000 | 129 | 69000 | 92.0 | | |
| $2 1/2$ | 6:1 | 140000 | 149 | 80000 | 107 | | |
| $2 5/8$ | 6:1 | 162000 | 168 | 90000 | 120 | | |
| $2 7/8$ | 6:1 | 180000 | 189 | 101000 | 137 | | |
| 3 | 6:1 | 200000 | 210 | 114000 | 153 | | |
| $3 1/4$ | 6:1 | 250000 | 264 | 137000 | 190 | | |
| $3 1/2$ | 6:1 | 300000 | 312 | 162000 | 232 | | |
| 4 | 6:1 | 360000 | 380 | 190000 | 276 | | |

Lbs/foot = Rope weight per linear foot
Break Lbs = Tensile strength
Safe Load Ratio = Break strength to safe load
Example: $7/16$" nylon rope break strength = 5000 lbs.
5000/10 = 500 lbs. safe working load

**Note: increased temperatures decrease rope strength**

# SAFE LOADS FOR SHACKLES

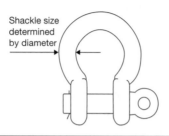

Shackle size determined by diameter

| Size (inches) | Safe Load at 90° (tons) |
|---|---|
| 1/4 | 1/3 |
| 5/16 | 1/2 |
| 3/8 | 3/4 |
| 7/16 | 1 |
| 1/2 | 1 1/2 |
| 5/8 | 2 |
| 3/4 | 3 |
| 7/8 | 4 |
| 1 | 5 1/2 |
| 1 1/8 | 6 1/2 |
| 1 1/4 | 8 |
| 1 3/8 | 10 |
| 1 1/2 | 12 |
| 1 3/4 | 16 |
| 2 | 21 |
| 2 1/4 | 27 |
| 2 1/2 | 34 |
| 2 3/4 | 40 |
| 3 | 50 |

# WIRE ROPE SLINGS — STRAIGHT LEG

| Rope Diameter (inches) | Straight 1 Leg | Choker 1 Leg | 60° Choker 2 Leg | 45° Choker 2 Leg | 30° Choker 2 Leg |
|---|---|---|---|---|---|
| | | | TONS CAPACITY | | |
| $1/4$ | $1/2$ | $1/3$ | $2/3$ | $1/2$ | $1/3$ |
| $3/8$ | 1 | $3/4$ | $1\frac{1}{4}$ | 1 | $3/4$ |
| $1/2$ | 2 | $1\frac{1}{2}$ | $2\frac{1}{2}$ | 2 | $1\frac{1}{2}$ |
| $5/8$ | 3 | 2 | 4 | 3 | 2 |
| $3/4$ | 4 | 3 | 5 | 4 | 3 |
| 1 | 7 | 5 | 8 | 7 | 5 |
| $1\frac{1}{4}$ | 10 | 7 | 12 | 9 | 7 |
| $1\frac{1}{2}$ | 13 | 9 | 16 | 13 | 9 |
| 2 | 21 | 15 | 27 | 22 | 15 |
| $2\frac{1}{2}$ | 28 | 22 | 38 | 31 | 22 |
| 3 | 36 | 28 | 49 | 40 | 28 |
| $3\frac{1}{2}$ | 40 | 34 | 59 | 48 | 34 |

# WIRE ROPE SLINGS – BASKET

| Rope Diameter (inches) | 60° Basket 2 Leg | 45° Basket 2 Leg | 30° Basket 2 Leg | 60° Basket 4 Leg | 45° Basket 4 Leg | 30° Basket 4 Leg |
|---|---|---|---|---|---|---|
| | | | TONS CAPACITY | | | |
| 1/4 | 2/3 | 1/2 | 1/3 | 1 | 1 | 3/4 |
| 3/8 | 1 1/2 | 1 | 3/4 | 3 | 2 | 1 1/2 |
| 1/2 | 2 1/2 | 2 | 1 1/2 | 5 | 4 | 3 |
| 5/8 | 4 | 3 | 2 | 7 | 6 | 4 |
| 3/4 | 5 | 4 | 3 | 11 | 9 | 6 |
| 1 | 9 | 7 | 5 | 18 | 15 | 10 |
| 1 1/4 | 13 | 11 | 7 | 26 | 21 | 15 |
| 1 1/2 | 17 | 14 | 10 | 35 | 28 | 20 |
| 2 | 27 | 22 | 15 | 53 | 44 | 31 |
| 2 1/2 | 38 | 31 | 22 | 75 | 61 | 43 |
| 3 | 49 | 40 | 29 | 97 | 80 | 56 |
| 3 1/2 | 59 | 49 | 34 | 118 | 97 | 68 |

## WIRE ROPE CHARACTERISTICS
## FOR 6 STRAND BY 19 WIRE TYPE

| WR Diameter (in.) | Weight (lbs./foot) | Breaking Point (lbs.) | Safe Load (lbs.) |
|---|---|---|---|
| $\frac{1}{4}$ | 0.10 | 4800 | 675 |
| $\frac{5}{16}$ | 0.16 | 7400 | 1000 |
| $\frac{3}{8}$ | 0.23 | 10600 | 1500 |
| $\frac{7}{16}$ | 0.31 | 14400 | 2000 |
| $\frac{1}{2}$ | 0.40 | 18700 | 2400 |
| $\frac{9}{16}$ | 0.51 | 23600 | 3300 |
| $\frac{5}{8}$ | 0.63 | 29000 | 4000 |
| $\frac{3}{4}$ | 0.90 | 41400 | 6000 |
| $\frac{7}{8}$ | 1.23 | 56000 | 8000 |
| 1 | 1.60 | 72800 | 10000 |
| $1\frac{1}{8}$ | 2.03 | 91400 | 13000 |
| $1\frac{1}{4}$ | 2.50 | 112400 | 16000 |
| $1\frac{3}{8}$ | 3.03 | 135000 | 19000 |
| $1\frac{1}{2}$ | 3.60 | 160000 | 22000 |
| $1\frac{3}{4}$ | 4.90 | 216000 | 30500 |
| 2 | 6.40 | 278000 | 40000 |
| $2\frac{1}{2}$ | 10.00 | 424000 | 60000 |

**The above values are for vertical pulls
at average ambient temperatures.**

## CABLE CLAMPS FOR WIRE ROPE

| Rope Diameter (inches) | Number of Clamps Required | Clip Spacing (inches) | Rope Turn-back (inches) |
|---|---|---|---|
| $\frac{1}{8}$ | 2 | 3 | $3\frac{1}{4}$ |
| $\frac{3}{16}$ | 2 | 3 | $3\frac{3}{4}$ |
| $\frac{1}{4}$ | 2 | $3\frac{1}{4}$ | $4\frac{3}{4}$ |
| $\frac{5}{16}$ | 2 | $3\frac{1}{4}$ | $5\frac{1}{4}$ |
| $\frac{3}{8}$ | 2 | 4 | $6\frac{1}{2}$ |
| $\frac{7}{16}$ | 2 | $4\frac{1}{2}$ | 4 |
| $\frac{1}{2}$ | 3 | 5 | $11\frac{1}{2}$ |
| $\frac{9}{16}$ | 3 | $5\frac{1}{2}$ | 12 |
| $\frac{5}{8}$ | 3 | $5\frac{3}{4}$ | 12 |
| $\frac{3}{4}$ | 4 | $6\frac{3}{4}$ | 18 |
| $\frac{7}{8}$ | 4 | 8 | 19 |
| 1 | 5 | $8\frac{3}{4}$ | 26 |
| $1\frac{1}{8}$ | 6 | $9\frac{3}{4}$ | 34 |
| $1\frac{1}{4}$ | 6 | $10\frac{3}{4}$ | 37 |
| $1\frac{7}{16}$ | 7 | $11\frac{1}{2}$ | 44 |
| $1\frac{1}{2}$ | 7 | $12\frac{1}{2}$ | 48 |
| $1\frac{5}{8}$ | 7 | $13\frac{1}{4}$ | 51 |
| $1\frac{3}{4}$ | 7 | $14\frac{1}{2}$ | 53 |
| 2 | 8 | $16\frac{1}{2}$ | 71 |
| $2\frac{1}{4}$ | 8 | $16\frac{1}{2}$ | 73 |
| $2\frac{1}{2}$ | 9 | $17\frac{3}{4}$ | 84 |
| $2\frac{3}{4}$ | 10 | 18 | 100 |
| 3 | 10 | 18 | 106 |

# FORKLIFTS

Approximately 95,000 employees are injured (and 100 are fatally injured) every year while operating powered industrial trucks. Forklift turnover accounts for a significant number of these fatalities.

- Train and certify all operators to ensure that they can manipulate forklifts safely.
- Do not allow any employee under 18 years old to operate a forklift.
- Properly maintain haulage equipment, including tires.
- Do not modify or make attachments that affect the capacity and safe operation of the forklift without written approval from the forklift's manufacturer.
- Examine the forklift truck for defects before using.
- Follow safe operating procedures for picking up, moving, putting down and stacking loads.
- Drive safely—never exceed 5 mph—and slow down in congested or slippery surface areas.
- Prohibit stunt driving and horseplay.
- Do not handle loads that are heavier than the capacity of the industrial truck.
- Remove unsafe or defective forklift trucks from service.
- Enforce seatbelt use for construction.
- Avoid traveling with elevated loads.
- Ensure that rollover protective structure is in place.
- Make certain that the reverse signal alarm is operational and audible above the surrounding noise level.

## INDOOR MINIMUM ILLUMINATION INTENSITIES IN FOOT-CANDLES

| Foot-Candles | Area of Operation |
|:---:|---|
| 5 | General construction area lighting. |
| 3 | General construction areas, concrete placement, excavation and waste areas, access ways, active storage areas, loading platforms and refueling, and field maintenance areas. |
| 5 | Indoors: warehouses, corridors, hallways, and exitways. |
| 5 | Tunnels, shafts and general underground work areas. (Exception: minimum of 10 foot-candles is required at tunnel and shaft heading during drilling, mucking, and scaling. Bureau of Mines–approved cap lights shall be acceptable for use in the tunnel heading.) |
| 10 | General construction plant and shops (e.g., batch plants, screening plants, mechanical and electrical equipment rooms, carpenter shops, rigging lofts and active store rooms, mess halls and indoor toilets, and workrooms). |
| 30 | First aid stations, infirmaries and offices. |

## SAFETY — PIPE LEGENDS

| Safety Classification | Color of Band | Color of Letters |
|---|---|---|
| Fire protection | Red | White |
| Dangerous | Yellow | Black |
| Safe | Green | Black |
| Protective | Blue | White |

| Outside Diameter of Pipe or Covering | Width of Color Band | Size of Legend Letters |
|---|---|---|
| ¾" to 1¼" | 8" | ½" |
| 1½" to 2" | 8" | ¾" |
| 2¼" to 6" | 12" | 1¼" |
| 8" to 10" | 24" | 2½" |
| Over 10" | 32" | 3½" |

# TYPES OF FIRE EXTINGUISHERS

Today they are virtually standard equipment in a business or residence and are rated by the makeup of the fire they will extinguish.

**TYPE A:** To extinguish fires involving trash, cloth, paper and other wood- or pulp-based materials. The flames are put out by water-based ingredients or dry chemicals.

**TYPE B:** To extinguish fires involving greases, paints, solvents, gas and other petroleum-based liquids. The flames are put out by cutting off oxygen and stopping the release of flammable vapors. Dry chemicals, foams and halon are used.

**TYPE C:** To extinguish fires involving electricity. The combustion is put out the same way as with a type B extinguisher but, most importantly, the chemical in a type C <u>MUST</u> be nonconductive to electricity in order to be safe and effective.

**TYPE D:** To extinguish fires involving combustible metals. Please be advised to obtain important information from your local fire department on the requirements for type D fire extinguishers for your area.

Any combination of letters indicate that an extinguisher will put out more than one type of fire. A type BC will put out two types of fires. The size of the fire to be extinguished is shown by a number in front of the letter, such as 100A. The following formulas apply:

**Class 1A** will extinguish 25 burning sticks 40" long.

**Class 1B** will extinguish a paint thinner fire 2.5 square feet in size.

A 100B fire extinguisher will put out a fire 100 times larger than a type 1B.

## Basic guidelines to follow:
- By using a type ABC, you will cover most basic fires.
- Use fire extinguishers with a gauge and ones that are constructed with metal. Also note whether the unit is Underwriter Laboratories approved.
- Utilize more than one extinguisher and be sure that each unit is mounted in a clearly visible and accessible manner.
- After purchasing any fire extinguisher, always review the basic instructions for its intended use. Never deviate from the manufacturer's guidelines. Following this simple procedure could save lives.

# FIRE-RESISTIVE LABELED EQUIPMENT

| Product Classification | SMNA Spec. | SMNA Class | UL Equiv. | Product Design and Test Features |
|---|---|---|---|---|
| Fire-Insulated Safe | F 1-D | A | A | 4-Hour-Tested Fire-Resistive Safe (with impact test) |
| Fire-Insulated Safe | F 1-D | B | B | 2-Hour-Tested Fire-Resistive Safe (with impact test) |
| Fire-Insulated Safe | F 1-D | C | C | 1-Hour-Tested Fire-Resistive Safe (with impact test) |
| Fire-Insulated Record Container | F 1-D | C | C | 1-Hour-Tested Fire-Resistive Container (with impact test) |
| Fire-Insulated Safe | F 1-ND | D | D | 1-Hour-Tested Fire-Resistive Safe (without impact test) |
| Fire-Insulated Ledger Tray | F 1-D | C | C | 1-Hour-Tested Fire-Resistive Ledger Tray (with impact test) |
| Fire-Insulated Container | F 2-ND | E | E | ½-Hour-Tested Fire-Resistive Container (without impact test) |
| Fire-Insulated Container | F 2-ND | D | D | 1-Hour-Tested Fire-Resistive Container (without impact test) |

| | | | | |
|---|---|---|---|---|
| Fire-Insulated Container | F 2-ND | 2 Hour | B | 2-Hour-Tested Fire-Resistive Container (without impact test) |
| Fire-Insulated Vault Door | F 3 | 2 Hour | 2 Hour | 2-Hour-Tested Fire-Resistive Vault Door |
| Fire-Insulated Vault Door | F 3 | 4 Hour | 4 Hour | 4-Hour-Tested Fire-Resistive Vault Door |
| Fire-Insulated Vault Door | F 3 | 6 Hour | 6 Hour | 6-Hour-Tested Fire-Resistive Vault Door |
| Fire-Insulated File Room Door | F 4 | 1 Hour | 1 Hour | 1-Hour-Tested Fire-Resistive File or Storage Room Door |
| Fire-Insulated Record Container Data Processing Safe | F 2-D | Class 150 | Class 150 | 2-Hour or 4-Hour-Tested Fire-Resistive Data Processing Safe |

**Class A**  Protects paper records from damage by fire (2,000°F) up to 4 hours.
**Class B**  Protects paper records from damage by fire (1,850°F) up to 2 hours.
**Classes C and D**  Protects paper records from damage by fire (1,700°F) up to 1 hour.
**Class E**  Protects paper records from damage by fire (1,550°F) up to ½ hour.
**Class 150**  Protects Electronic Data Processing records from damage by fire and humidity for a rated period.
**The Drop (or Impact) Test:**  Used to determine whether the fire-resistance of a product would be impaired if dropped 30' while still hot. Fire-resistant equipment is designed specifically to resist fire, and consists of a metal shell filled with fire-resistant insulation.

# FIRE RETARDATION RATINGS

## Partitions—Wood Framing (load-bearing)

| | Fire Rating | Ref. | Description |
|---|---|---|---|
| **Single Layer** | 45-min. | UL Design No.1 – 45 min. | 1/2" fire-shield gypsum wallboard, nailed both sides 2" × 4" studs, 16" o.c. |
| | 1-hour | UL Design No. 5 – 1 hr. | 5/8" fire-shield gypsum wallboard or fire-shield M-R board nailed both sides 2" × 4" wood studs, 16" o.c. |
| | 1-hour | UL Design No. 25 – 1 hr. | 5/8" fire-shield gypsum wallboard nailed both sides 2" × 4" wood studs, 24" o.c. |
| | 1-hour | FM Design WP-90 – 1 hr. | 5/8" fire-shield monolithic Durasan, vertically applied to 2" × 4" studs spaced 24" o.c. secured at joints with 6d nails spaced 7" o.c. and at intermediate studs with 3/8" × 3/8" bead of MC adhesive. |
| **Single Layer (resilient)** | 1-hour | Based on OSU T-3376 & UL Design No. 5 – 1 hr. | 5/8" fire-shield gypsum wallboard, screw applied to resilient furring channel, spaced 24" o.c. one side only, on 2" × 4" studs spaced 16" o.c. Other side 5/8" fire-shield gypsum wallboard nailed direct to studs. |
| | 1-hour | Based on OSU T-3376 | 5/8" fire-shield gypsum wallboard, screw applied to resilient furring channel, spaced 24" o.c. one side only, on 2" × 4" studs spaced 16" o.c. Other side 5/8" fire-shield gypsum wallboard screw attached at 16" spacing, 3" fiberglass in stud cavity. |
| **Single Layer** | 1-hour | OSU T-3376 | 5/8" fire-shield gypsum wallboard, screw applied to resilient furring channels 24" o.c. nailed to both sides of 2" × 4" studs spaced 16" o.c. |

1-32

| Category | Rating | Design | Description |
|---|---|---|---|
| Double Layer | 1-hour | FM Design WP-147 – 1 hr. | 1/2" fire-shield wallboard or Durasan laminated to 1/4" gypsum wallboard nailed to both sides 2" × 4" studs spaced 16" o.c. |
| | 2-hour | Based on UL Design No. 4 – 2 hr. | 5/8" fire-shield gypsum wallboard base layer nail applied to 2" × 4" wood stud spaced 16" o.c. Face layer 5/8" fire-shield gypsum wallboard laminated and nail applied. |
| Exterior Walls | 1-hour | FM Design WP-78 – 1 hr. UL Design | 5/8" fire-shield gypsum wallboard nailed horizontally to inside face of 2" × 4" wood studs 16" o.c.; 1/2" gypsum sheathing nailed to outside face of studs. Siding 3/8" woodrock. |
| | 2-hour | No. 23 – 2 hr. | Two layers 5/8" fire-shield gypsum wallboard nailed horiz. or vert. to inside face of 2" × 4" wood studs 16" o.c.; 1/2" gypsum sheathing nailed to outside face of studs, brick veneer facing. |
| Double Layer (with Deciban Sound Deadening Board) | Non-rated | — | 1/2" Deciban nail applied both sides 2" × 4" wood studs 16" o.c. Face layer 1/2" gypsum wallboard laminated. |
| | Non-rated | — | 1/2" Deciban nail applied to both sides 2" × 4" wood studs, 16" o.c. fire-topped face layer. 5/8" gypsum wallboard laminated. |
| | 1-hour | UL Design No. 17 – 1 hr. | 1/2" Deciban nail applied to both sides 2" × 3" wood studs, staggered 16" o.c. on 2" × 3" plates spaced 1" apart. Face layer 5/8" fire-shield gypsum wallboard nail applied. |
| | 1-hour | UL Design No. 26 – 1 hr. | 1/2" Deciban nail applied both sides 2" × 3" wood studs, staggered 24" o.c. on 2" × 3" plates spaced 1" apart. Face layer 1/2" fire-shield gypsum wallboard nail applied. |

# FIRE RETARDATION RATINGS (cont.)

## Partitions—Steel Framing

| | Fire Rating | Ref. | Description |
|---|---|---|---|
| **Unbalanced 2½" Studs** | 1-hour | FM Design WP-66 | ½" fire-shield (Monolithic Durasan) vertically applied to 2½" screw stud. Double layer one side, single layer on the other. Base layer screw attached, face layer and single layer screwed at edges, adhesively attached along center. |
| | 1-hour | Based on FM Design WP-66 | ½" fire-shield gypsum wallboard screw attached vertically to both sides, 2½" screw studs spaced 24" o.c. Second layer screw attached vertically to one side only. |
| | 1-hour | Based on FM Design WP-66 | ½" fire-shield gypsum wallboard screw attached vertically to both sides, 2½" screw studs spaced 24" o.c. Second layer screw attached vertically to one side only and 3" fiberglass in cavity. |
| | 1½-hour | Based on OSU T-3240 | ⅝" fire-shield gypsum wallboard screw attached vertically to both sides, 2½" screw studs spaced 24" o.c. Second layer screw attached vertically to one side only. |
| | 1½-hour | Based on OSU T-3240 | ⅝" fire-shield gypsum wallboard screw attached vertically to both sides, 2½" screw studs spaced 24" o.c. Second layer screw attached vertically to one side only and 3" fiberglass in cavity. |
| **3⅝" Studs** | 1-hour | Based on FM Design WP-66 | ½" fire-shield gypsum wallboard screw attached vertically to both sides, 3⅝" screw studs spaced 24" o.c. Second layer screw attached vertically to one side only. |
| | 1-hour | Based on FM Design WP-66 | ½" fire-shield gypsum wallboard screw attached vertically to both sides, 3⅝" screw studs spaced 24" o.c. Second layer screw attached vertically to one side only and 3" fiberglass in cavity. |
| | 1½-hour | OSU T-3240 | ⅝" fire-shield gypsum wallboard screw attached vertically to both sides, 3⅝" screw studs spaced 24" o.c. Second layer laminated vertically and screwed to one side only. |

| Studs | Rating | Design | Description |
|---|---|---|---|
| 1⅝" Studs — Single Layer | 1-hour | OSU T-3296 | ⅝" fire-shield gypsum wallboard screw attached vertically to both sides, 1⅝" screw studs 24" o.c. |
| 1⅝" Studs — Single Layer | 1-hour | Based on OSU T-3296 | ⅝" fire-shield gypsum wallboard screw attached vertically to both sides, 1⅝" screw studs 24" o.c. with 1" fiberglass in cavity. |
| 2½" Studs — Single Layer | 1-hour | Based on OSU T-3296 | ⅝" fire-shield gypsum wallboard screw attached vertically to both sides, 2½" screw studs 24" o.c. |
| 2½" Studs — Single Layer | 1-hour | Based on OSU T-3296 | ⅝" fire-shield gypsum wallboard screw attached vertically to both sides, 2½" screw studs 24" o.c. with 3" fiberglass in cavity. |
| 2½" Studs — Single Layer | 1-hour | FM Design WP-51 | ½" fire-shield gypsum wallboard screw attached vertically to both sides, 2½" screw studs 24" o.c., 2" mineral wool in stud cavity. |
| 3⅝" Studs — Single Layer | 1-hour | FM Design WP-45 | ⅝" fire-shield gypsum wallboard screw attached horizontally to both sides, 3⅝" screw studs 24" o.c. Wallboard joints staggered. |
| 3⅝" Studs — Single Layer | 1-hour | Based on OSU T-1770 | ⅝" fire-shield gypsum wallboard screw attached vertically to both sides, 3⅝" screw studs 24" o.c. with 3" fiberglass in cavity. |
| 3⅝" Studs — Single Layer | 45-min. | Based on FM Design WP-51 | ½" fire-shield gypsum wallboard screw attached vertically to both sides, 3⅝" screw studs 24" o.c., 2" fiberglass in cavity. |
| 3⅝" Studs — Single Layer | 1-hour | Based on FM Design WP-51 | ½" fire-shield gypsum wallboard screw attached vertically to both sides, 3⅝" screw studs 24" o.c., 2" mineral wool in stud cavity. |
| 3⅝" Studs — Single Layer | 1-hour | OSU T-1770 | ⅝" fire-shield gypsum wallboard screw attached vertically to both sides, 3⅝" screw studs 24" o.c. |

# FIRE RETARDATION RATINGS *(cont.)*

## Partitions—Steel Framing *(cont.)*

| | | Fire Rating | Ref. | Description |
|---|---|---|---|---|
| **3⅝" Studs**<br>**Unbalanced** | | 1½-hour | Based on OSU T-3240 | ⅝" fire-shield gypsum wallboard screw attached vertically to both sides, 3⅝" screw studs 24" o.c. Second layer laminated vertically and screwed to one side only and 2" fiberglass or min. wool in cavity. |
| **2½" Studs**<br>**Double Layer** | | 1-hour | FM Design WP-152 | ½" fire-shield wallboard or Durasan laminated to ¼" wallboard screw attached both sides, 2½" screw studs spaced 24" o.c. |
| | | 1-hour | Based on FM Design WP-152 | ½" fire-shield wallboard or Durasan laminated to ¼" wallboard screw attached both sides, 2½" screw studs spaced 24" o.c. with 2" fiberglass in cavity. |
| | | 2-hour | OSU T-3370 | Two layers ½" fire-shield gypsum wallboard screw attached vertically both sides, 2½" screw studs spaced 24" o.c. Vertical joints staggered. |
| | | 2-hour | FM Design WP-47 | First layer ½" fire-shield gypsum wallboard screw attached vertically both sides, 2½" screw studs spaced 24" o.c. Second layer screw attached horizontally both sides. |
| | | 2-hour | Based on OSU T-3370 | Two layers ½" fire-shield gypsum wallboard screw attached vertically both sides, 2½" screw studs spaced 24" o.c. Vertical joints staggered and 3" fiberglass in cavity. |

1-36

| | | | |
|---|---|---|---|
| **Double Layer** **3⁵⁄₈" Studs** | 2-hour | Based on FM Design WP-47 OSU T-1771 | First layer ⁵⁄₈" fire-shield gypsum wallboard screw attached vertically both sides, 2¹⁄₂" screw studs spaced 24" o.c. Second layer screw attached horizontally both sides and 3" fiberglass in cavity. |
| | 2-hour | Based on OSU T-3370 | Two layers ¹⁄₂" fire-shield gypsum wallboard screw attached vertically both sides, 3⁵⁄₈" screw studs spaced 24" o.c. Vertical joints staggered. |
| | 2-hour | Based on OSU T-3370 | Two layers ¹⁄₂" fire-shield gypsum wallboard screw attached vertically both sides, 3⁵⁄₈" screw studs spaced 24" o.c. Vertical joints staggered and 3" fiberglass in cavity. |
| | 2-hour | OSU T-1771 | First layer ⁵⁄₈" fire-shield gypsum wallboard screw attached vertically both sides, 3⁵⁄₈" screw studs spaced 24" o.c. Second layer screw laminated vertically both sides. |
| | 2-hour | Based on FM Design WP-47 | First layer ⁵⁄₈" fire-shield gypsum wallboard screw attached vertically both sides, 3⁵⁄₈" screw studs spaced 24" o.c. Second layer screw attached horizontally both sides and 3" fiberglass in cavity. |

# CLASSIFICATION OF SOFTWOOD PLYWOODS RATES SPECIES FOR STRENGTH AND STIFFNESS

## Group 1

| | | | |
|---|---|---|---|
| Apitong<br>Beech,<br>  American<br>Birch,<br>  Sweet<br>  Yellow | Douglas<br>  Fir 1<br>Kapur<br>Keruing<br>Larch,<br>  Western | Maple,<br>  Sugar<br>Pine,<br>  Caribbean<br>  Ocote | Pine, South<br>  Loblolly<br>  Longleaf<br>  Shortleaf<br>  Slash<br>Tanoak |

## Group 2

| | | | |
|---|---|---|---|
| Cedar, Port<br>  Orford<br>Cypress<br>Douglas<br>  Fir 2<br>Fir,<br>  California<br>  Red<br>  Grand<br>  Noble | Fir *(cont.)*<br>  Pacific<br>  Silver<br>  White<br>Hemlock,<br>  Western<br>Lauan,<br>  Almon<br>  Bagtikan<br>  Mayapis<br>  Red Lauan | Lauan *(cont.)*<br>  Tangile<br>  White<br>Maple, Black<br>Mengkulang<br>Meranti, Red<br>Mersawa<br>Pine,<br>  Pond<br>  Red | Pine *(cont.)*<br>  Virginia<br>  Western<br>    White<br>Spruce,<br>  Red<br>  Sitka<br>Sweetgum<br>Tamarack<br>Yellow-poplar |

## Group 3

| | | | |
|---|---|---|---|
| Adler, Red<br>Birch,<br>  Paper<br>Cedar,<br>  Alaska | Fir, Subalpine<br>Hemlock,<br>  Eastern<br>Maple,<br>  Bigleaf | Pine,<br>  Jack<br>  Lodgepole<br>  Ponderosa<br>  Spruce | Redwood<br>Spruce,<br>  Black<br>  Engelmann<br>  White |

## Group 4

| | | | |
|---|---|---|---|
| Aspen,<br>  Bigtooth<br>  Quaking<br>Cativo | Cedar,<br>  Incense<br>  Western Red<br>Cottonwood,<br>  Eastern | Cottonwood<br>*(cont.)*<br>  Black<br>  (Western<br>    Poplar) | Pine,<br>  Eastern<br>    White<br>  Sugar |

## Group 5

| | | | |
|---|---|---|---|
| Basswood | Fir,<br>  Balsam | Poplar,<br>  Balsam | |

**Note:** Group 1 represents strongest woods.

## PLYWOOD MARKING

**APA**

Panel grade —————— RATED SHEATHING

Span rating —————— **32/16** 1/2 INCH ——— Thickness

SIZED FOR SPACING

Exposure durability ——— EXPOSURE 1
classification

—————— Mill number

**000**

National Research ——— NRB-108
Board report number

**Typical-grade trademark that is stamped on all plywood manufactured in compliance with panels national plywood standard.**

1-39

# PLYWOOD GRADES AND USAGE

## Protected or Interior Use

| Grade Designation | Description & Common Uses | Typical Trademarks |
|---|---|---|
| APA Rated Sturd-I-Floor EXP 1 or 2 | For combination subfloor-underlayment. Provides smooth surface for application of resilient floor covering and possesses high concentrated and impact load resistance. Can be manufactured as conventional veneered plywood, a composite or a nonveneered panel. Available square edge or tongue-and-groove. Specify Exposure 1 when long construction delays are anticipated. Common thicknesses: $\frac{5}{8}$" ($\frac{19}{32}$"), $\frac{3}{4}$" ($\frac{23}{32}$"). | **APA** RATED STURD-I-FLOOR SIZED FOR SPACING **24 oc** 23/32 INCH T&G NET WIDTH 47-1/2 EXPOSURE 1 000 NRB-108 |
| APA Rated Sturd-I-Floor 48 oc (2-4-1) EXP 1 | For combination subfloor-underlayment on 32" and 48" spans and for heavy timber roof construction. Provides smooth surface for application of resilient floor coverings and possesses high concentrated and impact load resistance. Manufactured only as conventional veneered plywood and only with exterior glue (Exposure 1). Available square edge or tongue-and-groove. Thickness: $1\frac{1}{8}$" . | **APA** RATED STURD-I-FLOOR **48 oc** 1-1/8 INCH SIZED FOR SPACING (2-4-1) EXPOSURE 1 T&G 000 INT/EXT GLUE NRB-108 FHA-UM-66 |

## Exterior Use

| | | |
|---|---|---|
| APA Rated Sheathing EXT | Exterior sheathing panel for subflooring and wall and roof sheathing, siding on service and farm buildings, crating, pallets, pallet bins, cable reels, etc. Manufactured as conventional veneered plywood. Common thicknesses: $5/16''$, $3/8''$, $1/2''$, $5/8''$, $3/4''$. | **APA** RATED SHEATHING **48/24** 3/4 INCH SIZED FOR SPACING EXTERIOR 000 NRB-108 |
| APA Structural I & II Rated Sheathing EXT | For engineered applications in construction and industry where resistance to weather or moisture is required. Manufactured only as conventional veneered PS1 plywood. Structural I more commonly available. Common thicknesses: $5/16''$, $3/8''$, $1/2''$, $5/8''$, $3/4''$. | **APA** RATED SHEATHING STRUCTURAL I **42/20** 5/8 INCH SIZED FOR SPACING EXTERIOR 000 PS 1-74 C-C NRB-108 |
| APA Rated Sturd-I-Floor EXT | For combination subfloor-underlayment under resilient floor coverings where severe moisture conditions may be present, as in balcony decks. Possesses high concentrated and impact load resistance. Manufactured only as conventional veneered plywood. Available square edge or tongue-and-groove. Common thicknesses: $5/8''$ ($19/32''$), $3/4''$ ($23/32''$). | **APA** RATED STURD-I-FLOOR **20 OC** 19/32 INCH SIZED FOR SPACING EXTERIOR 000 NRB-108 |

**Note:** Specific grades, thicknesses, constructions and exposure durability classifications may be in limited supply in some areas. Check with your supplier before specifying. Specify performance-rated panels by thickness and span rating.

# PLYWOOD GRADES AND USAGE (cont.)

## Protected or Interior Use

| Grade Designation | Description & Common Uses | Typical Trademarks |
|---|---|---|
| APA Rated Sheathing EXP 1 or 2 | Specially designed for subflooring and wall and roof sheathing, but can also be used for a broad range of other construction and industrial applications. Can be manufactured as conventional veneered plywood, a composite or a nonveneered panel. For special engineered applications, including high load requirements and certain industrial uses, veneered panels conforming to PS 1 may be required. Specify Exposure 1 when long construction delays are anticipated. Common thicknesses: $5/16"$, $3/8"$, $7/16"$, $1/2"$, $5/8"$, $3/4"$. | APA RATED SHEATHING 32/16 1/2 INCH SIZED FOR SPACING EXPOSURE 1 000 NRB-108 |
| APA Structural I & II Rated Sheathing EXP 1 | Unsanded all-veneer PS 1 plywood grades for use where strength properties are of maximum importance: structural diaphragms, box beams, gusset plates, stressed-skin panels, containers, pallet bins. Made only with exterior glue (Exposure 1). STRUCTURAL I more commonly available. Common thicknesses: $5/16"$, $3/8"$, $1/2"$, $5/8"$, $3/4"$. | APA RATED SHEATHING STRUCTURAL I 24/0 3/8 INCH SIZED FOR SPACING EXPOSURE 1 000 PS 1-74 C-D INT/EXT GLUE NRB-108 |

## SOFTWOOD PLYWOOD VENEER GRADES

| | |
|---|---|
| **N** | Smooth surface "natural finish" veneer. Select, all heartwood or all sapwood. Free of open defects. Allows not more than six repairs, wood only, per 4 × 8 panel, made parallel to grain and well matched for grain and color. |
| **A** | Smooth, paintable. Not more than 18 neatly made repairs, boat, sled, or router type, and parallel to grain, permitted. May be used for natural finish in less demanding applications. |
| **B** | Solid surface. Shims, circular repair plugs and tight knots to 1" across grain permitted. Some minor splits permitted. |
| **C Plugged** | Improved C veneer with splits limited to $\frac{1}{8}$" width and knotholes and borer holes limited to $\frac{1}{4}$" × $\frac{1}{2}$". Admits some broken grain. Synthetic repairs permitted. |
| **C** | Tight knots to $1\frac{1}{2}$". Knotholes to 1" across grain, and some to $1\frac{1}{2}$" if total width of knots and knotholes is within specified limits. Synthetic or wood repairs. Discoloration and sanding defects that do not impair strength permitted. Limited splits allowed. Stitching permitted. |
| **D** | Knots and knotholes to $2\frac{1}{2}$" width across grain and $\frac{1}{2}$" larger within specified limits. Limited splits allowed. Stitching permitted. Limited to Interior, Exposure 1 and Exposure 2 panels. |

## EXPOSED PLYWOOD PANEL SIDING

| Minimum Thickness[1] | Minimum No. of Plies | Stud Spacing (inches) Plywood Siding Applied Direct to Studs or Over Sheathing |
|:---:|:---:|:---:|
| $3/8$" | 3 | 16[1] |
| $1/2$" | 4 | 24 |

Thickness of grooved panels is measured at bottom of grooves.
[1]May be 24" if plywood siding applied with face grain perpendicular to studs or over one of the following:
a) 1" board sheathing.
b) $15/32$" plywood sheathing.
c) $3/8$" plywood sheathing with face grain of sheathing perpendicular to studs.

## ALLOWABLE SPANS FOR EXPOSED PARTICLEBOARD PANEL SIDING

| Grade | Stud Spacing (inches) | Minimum Thickness (inches) | | Exterior Ceilings and Soffits |
| | | Siding | | |
| | | Direct to Studs | Continuous Support | Direct to Supports |
|:---|:---:|:---:|:---:|:---:|
| 2-M-W | 16 | $3/8$ | $5/16$ | $5/16$ |
| | 24 | $1/2$ | $5/16$ | $3/8$ |
| 2-M-1 | 16 | $5/8$ | $3/8$ | — |
| 2-M-2 | — | — | — | — |
| 2-M-3 | 24 | $3/4$ | $3/8$ | — |

## ALLOWABLE SPANS FOR PARTICLEBOARD SUBFLOOR AND COMBINED SUBFLOOR-UNDERLAYMENT

| Grade | Thickness (inches) | Maximum Spacing of Supports | |
|---|---|---|---|
| | | Subfloor | Combined Subfloor-Underlayment |
| 2-M-W | $1/2$ | 16 | — |
| | $5/8$ | 20 | 16 |
| | $3/4$ | 24 | 24 |
| 2-M-3 | $3/4$ | 20 | 20 |

**Note:** All panels are continuous over two or more spans. Uniform deflection limitation: $1/360$ of the span under 100 per square foot minimum load. Edges shall have tongue-and-groove joints or shall be supported with blocking. The tongue-and-groove panels are installed with the long dimension perpendicular to supports.

## ALLOWABLE SPANS FOR PLYWOOD COMBINATION SUBFLOOR-UNDERLAYMENT

### Plywood Continuous Over Two or More Spans and Face Grain Perpendicular to Supports

| Identification | Spacing of Joists (inches) | | | |
|---|---|---|---|---|
| | 16 | 20 | 24 | 48 |
| Species Group[1] | Thickness (inches) | | | |
| 1 | $1/2$ | $5/8$ | $3/4$ | — |
| 2, 3 | $5/8$ | $3/4$ | $7/8$ | — |
| 4 | $3/4$ | $7/8$ | 1 | — |
| Span Rating[2] | 16 o.c. | 20 o.c. | 24 o.c. | 48 o.c. |

Spans limited to value shown because of possible effect of concentrated loads. Allowable uniform load based on deflection of $1/360$ of span is 125 psf, except allowable total uniform load for $1 1/8$" plywood over joists spaced 48" on center is 65 psf. Plywood edges shall have approved tongue-and-groove joints or shall be supported with blocking, unless $1/4$" minimum thickness underlayment is installed, or finish floor is $3/4$" wood strip. If wood strips are perpendicular to supports, thicknesses shown for 16" and 20" spans may be used on 24" span.

[1]Applicable to all grades of sanded exterior-type plywood.
[2]Applicable to underlayment grade and C-C (plugged).

## LOAD LIMITS FOR WOOD GIRDERS

| Wood Girders | Safe Load in lb. for Spans From 6' to 10' | | | | |
|---|---|---|---|---|---|
| Size | 6' | 7' | 8' | 9' | 10' |
| 6 × 8 Solid | 8,306 | 7,118 | 6,220 | 5,539 | 4,583 |
| 6 × 8 Built-up | 7,359 | 6,306 | 5,511 | 4,908 | 4,062 |
| 6 × 10 Solid | 11,357 | 10,804 | 9,980 | 8,887 | 7,997 |
| 6 × 10 Built-up | 10,068 | 9,576 | 8,844 | 7,878 | 7,086 |
| 8 × 8 Solid | 11,326 | 9,706 | 8,482 | 7,553 | 6,250 |
| 8 × 8 Built-up | 9,812 | 8,408 | 7,348 | 6,544 | 5,416 |
| 8 × 10 Solid | 15,487 | 14,782 | 13,608 | 12,116 | 10,902 |
| 8 × 10 Built-up | 13,424 | 12,768 | 11,792 | 10,504 | 9,448 |

# LOADING TABLES FOR FLOOR/CEILING TRUSSES BUILT WITH 2 × 4 CHORDS

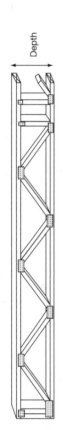

Depth

## Douglas Fir #1 Dense MC 15

### Depth (inches)

| Span (feet) | 12.0 | 14.0 | 16.0 | 18.0 | 20.0 | 22.0 | 24.0 | 26.0 | 28.0 | 30.0 |
|---|---|---|---|---|---|---|---|---|---|---|
| | | | | | | | | | | Maximum load* |
| 16.0 | 171.2 | 196.6 | 216.9 | 235.3 | 252.0 | 267.3 | 281.4 | 294.4 | 306.4 | 317.5 |
| 18.0 | 124.7 | 166.6 | 185.2 | 202.2 | 218.0 | 232.6 | 246.1 | 258.8 | 270.5 | 281.5 |
| 20.0 | 95.0 | 128.4 | 159.1 | 174.8 | 189.4 | 203.1 | 215.9 | 227.9 | 239.3 | 249.9 |
| 22.0 | 75.1 | 100.2 | 129.6 | 150.4 | 165.5 | 178.1 | 190.1 | 201.4 | 212.2 | 222.3 |
| 24.0 | 61.3 | 80.6 | 103.3 | 126.9 | 141.9 | 156.8 | 168.1 | 178.7 | 188.7 | 198.4 |
| 26.0 | 51.4 | 66.6 | 84.4 | 104.9 | 121.4 | 134.2 | 146.9 | 159.1 | 168.5 | 177.5 |
| 28.0 | 44.2 | 56.3 | 70.6 | 87.0 | 105.0 | 116.1 | 127.2 | 138.2 | 149.2 | 159.5 |
| 30.0 | 38.7 | 48.6 | 60.2 | 73.5 | 88.6 | 101.4 | 111.1 | 120.8 | 130.4 | 140.0 |
| 32.0 | 34.5 | 42.7 | 52.2 | 63.2 | 75.6 | 89.3 | 97.9 | 106.4 | 114.9 | 123.4 |
| 34.0 | 31.3 | 38.1 | 46.0 | 55.2 | 65.5 | 77.1 | 86.9 | 94.5 | 102.1 | 109.6 |

*Loads are shown in pounds per lineal foot. Deflection may govern the span limits.
These tables are limited to a simple span condition uniformly loaded.

# LOADING TABLES FOR FLOOR/CEILING TRUSSES BUILT WITH 2 × 4 CHORDS *(cont.)*

Depth

## Southern Pine #1 Dense KD

### Depth (inches)

| Span (feet) | 12.0 | 14.0 | 16.0 | 18.0 | 20.0 | 22.0 | 24.0 | 26.0 | 28.0 | 30.0 |
|---|---|---|---|---|---|---|---|---|---|---|
| | | | | | | | | | | Maximum load* |
| 16.0 | 164.2 | 193.3 | 212.8 | 230.5 | 246.5 | 261.1 | 274.4 | 286.7 | 298.0 | 308.5 |
| 18.0 | 119.8 | 163.5 | 182.2 | 198.7 | 213.9 | 227.9 | 240.9 | 252.9 | 264.1 | 274.6 |
| 20.0 | 91.4 | 123.2 | 153.9 | 172.2 | 186.4 | 199.6 | 211.9 | 223.4 | 234.3 | 244.5 |
| 22.0 | 72.4 | 96.3 | 124.4 | 145.0 | 162.0 | 175.5 | 187.0 | 198.0 | 208.3 | 218.1 |
| 24.0 | 59.2 | 77.6 | 99.3 | 122.4 | 136.8 | 151.2 | 165.5 | 176.0 | 185.7 | 195.0 |
| 26.0 | 49.8 | 64.3 | 81.3 | 100.8 | 117.0 | 129.4 | 141.6 | 153.9 | 166.0 | 174.9 |
| 28.0 | 42.8 | 54.4 | 68.1 | 83.7 | 101.2 | 111.9 | 122.6 | 133.2 | 143.8 | 154.3 |
| 30.0 | 37.6 | 47.1 | 58.2 | 70.9 | 85.2 | 97.8 | 107.1 | 116.4 | 125.7 | 134.9 |
| 32.0 | 33.6 | 41.4 | 50.6 | 61.0 | 72.9 | 86.1 | 94.4 | 102.6 | 110.8 | 119.0 |
| 34.0 | 30.5 | 37.0 | 44.6 | 53.4 | 63.3 | 74.3 | 83.8 | 91.1 | 98.4 | 105.7 |

*Loads are shown in pounds per lineal foot. Deflection may govern the span limits. These tables are limited to a simple span condition uniformly loaded.

1-48

# LOADING TABLES FOR FLOOR/CEILING TRUSSES BUILT WITH 2 × 4 CHORDS (cont.)

Depth

## Machine Rated Lumber 2100F-1.8E

|  |  |  |  |  |  | Maximum load* |  |  |  |  |
|---|---|---|---|---|---|---|---|---|---|---|
|  | | | | | Depth (inches) | | | | | |
| Span (feet) | 12.0 | 14.0 | 16.0 | 18.0 | 20.0 | 22.0 | 24.0 | 26.0 | 28.0 | 30.0 |
| 16.0 | 156.3 | 192.0 | 211.2 | 228.6 | 244.3 | 258.6 | 271.7 | 283.8 | 294.9 | 305.1 |
| 18.0 | 114.2 | 155.7 | 181.1 | 197.3 | 212.3 | 226.0 | 238.8 | 250.6 | 261.6 | 271.9 |
| 20.0 | 87.4 | 117.5 | 153.0 | 171.2 | 185.1 | 198.1 | 210.3 | 221.7 | 232.3 | 242.3 |
| 22.0 | 69.4 | 92.0 | 118.7 | 149.2 | 162.2 | 174.3 | 185.8 | 196.5 | 206.7 | 216.4 |
| 24.0 | 56.9 | 74.3 | 94.8 | 118.4 | 142.8 | 154.1 | 164.7 | 174.9 | 184.5 | 193.6 |
| 26.0 | 47.9 | 61.7 | 77.8 | 96.3 | 117.2 | 136.8 | 146.7 | 156.1 | 165.2 | 173.8 |
| 28.0 | 41.4 | 52.4 | 65.3 | 80.1 | 96.9 | 115.5 | 131.2 | 139.9 | 148.4 | 156.5 |
| 30.0 | 36.4 | 45.4 | 55.9 | 67.9 | 81.5 | 96.7 | 113.4 | 125.9 | 133.8 | 141.3 |
| 32.0 | 32.7 | 40.0 | 48.7 | 58.6 | 69.8 | 82.3 | 96.1 | 111.2 | 121.0 | 128.1 |
| 34.0 | 29.7 | 35.9 | 43.1 | 51.4 | 60.7 | 71.1 | 82.6 | 95.2 | 108.8 | 116.5 |

*Loads are shown in pounds per lineal foot. Deflection may govern the span limits. These tables are limited to a simple span condition uniformly loaded.

## ALLOWABLE UNIFORM LOADS FOR W STEEL BEAMS

| Designation (wt./ft.) | Nominal Size (dp. × wd.) | Span in Feet | | | | | | | | | | |
|---|---|---|---|---|---|---|---|---|---|---|---|---|
| | | 8 | 10 | 12 | 14 | 16 | 18 | 20 | 22 | 24 | 26 |
| W8 × 10 | 8 × 4 | 15.6 | 12.5 | 10.4 | 8.9 | 7.8 | 6.9 | – | – | – | – |
| W8 × 13 | 8 × 4 | 19.9 | 15.9 | 13.3 | 11.4 | 9.9 | 8.8 | – | – | – | – |
| W8 × 15 | 8 × 4 | 23.6 | 18.9 | 15.8 | 13.5 | 11.8 | 10.5 | – | – | – | – |
| W8 × 18 | 8 × 5¹/₄ | 30.4 | 24.3 | 20.3 | 17.4 | 15.2 | 13.5 | – | – | – | – |
| W8 × 21 | 8 × 5¹/₄ | 36.4 | 29.1 | 24.3 | 20.8 | 18.2 | 16.2 | – | – | – | – |
| W8 × 24 | 8 × 6¹/₂ | 41.8 | 33.4 | 27.8 | 23.9 | 20.9 | 18.6 | – | – | – | – |
| W8 × 28 | 8 × 6¹/₂ | 48.6 | 38.9 | 32.4 | 27.8 | 24.3 | 21.6 | – | – | – | – |
| W10 × 22 | 10 × 5³/₄ | – | – | 30.9 | 26.5 | 23.2 | 20.6 | 18.6 | 16.9 | – | – |
| W10 × 26 | 10 × 5³/₄ | – | – | 37.2 | 31.9 | 27.9 | 24.8 | 22.3 | 20.3 | – | – |
| W10 × 30 | 10 × 5³/₄ | – | – | 43.2 | 37.0 | 32.4 | 28.8 | 25.9 | 23.6 | – | – |
| W12 × 26 | 12 × 6¹/₂ | – | – | – | – | 33.4 | 29.7 | 26.7 | 24.3 | 22.3 | 20.5 |
| W12 x 30 | 12 × 6¹/₂ | – | – | – | – | 38.6 | 34.3 | 30.9 | 28.1 | 25.8 | 23.8 |
| W12 × 35 | 12 × 6¹/₂ | – | – | – | – | 45.6 | 40.6 | 36.5 | 33.2 | 30.4 | 28.1 |

**Note:** Loads are given in kips (1 kip = 1000 lb).

# OPTIMUM SPACING OF SPRING CLIPS FOR CEILING

## 25-Gauge Studs—Maximum Heights

| Stud Spacing | 1⅝" Stud | | 2½" Stud | | 3¼" Stud | | 3⅝" Stud | | 4" Stud | |
|---|---|---|---|---|---|---|---|---|---|---|
| | ½" GWB | ⅝" GWB | ½" GWB | ⅝" GWB | ½" GWB | ⅝" GWB | ½" GWB | ⅝" GWB | ½" GWB | ⅝" GWB |
| 12" o.c. | 12' 0" | 12' 4" | 16' 2" | 16' 6" | 19' 6" | 19' 10" | 21' 0" | 21' 4" | 22' 0" | 22' 4" |
| 16" o.c. | 11' 0" | 11' 7" | 14' 8" | 15' 5" | 17' 10" | 18' 4" | 19' 5" | 19' 11" | 20' 8" | 20' 10" |
| 24" o.c. | 10' 0" | 10' 10" | 13' 5" | 14' 3" | 15' 10" | 16' 7" | 17' 3" | 18' 2" | 18' 5" | 19' 2" |

## 20-Gauge Studs—Maximum Heights

| Stud Spacing | 2½" Stud | | 3¼" Stud | | 3⅝" Stud | | 4" Stud | |
|---|---|---|---|---|---|---|---|---|
| | ½" WB | ⅝" WB | ½" WB | ⅝" WB | ½" WB | ⅝" WB | ½" WB | ⅝" WB |
| 12" o.c. | 18' 9" | 19' 0" | 22' 8" | 22' 11" | 24' 8" | 24' 10" | 26' 7" | 26' 10" |
| 16" o.c. | 17' 9" | 18' 0" | 21' 7" | 21' 9" | 23' 5" | 23' 7" | 25' 3" | 25' 6" |
| 24" o.c. | 15' 9" | 16' 0" | 18' 11" | 19' 2" | 20' 5" | 20' 8" | 22' 0" | 22' 3" |

## HEATING TEMPERATURES OF ROOF ASPHALT

| ASTM D-312 Type No. | Asphalt Type | Maximum Heating Temperature (°F) |
|---|---|---|
| I | Dead Level | 475 |
| II | Flat | 500 |
| III | Steep | 525 |
| IV | Special steep | 525 |

## ROOF NAILER SPACING

| Incline (in.) | Smooth | Gravel | Cap Sheet |
|---|---|---|---|
| 0–1 | Not required | Not required | Not required |
| 1–2 | Not required | 20' face to face | 20' face to face |
| 2–3 | 20' face to face | 10' face to face | 10' face to face |
| 3–4 | 10' face to face | Not recommended | 4' face to face |
| 4–6 | 4' face to face | Not recommended | 4' face to face |

| TIGHTENING TORQUE IN POUND-FEET-SCREW FIT | | | |
|---|---|---|---|
| Wire Size, AWG/kcmil | Driver | Bolt | Other |
| 18-16 | 1.67 | 6.25 | 4.2 |
| 14-8 | 1.67 | 6.25 | 6.125 |
| 6-4 | 3.0 | 12.5 | 8.0 |
| 3-1 | 3.2 | 21.0 | 10.40 |
| 0-2/0 | 4.22 | 29 | 12.5 |
| 3/0-200 | – | 37.5 | 17.0 |
| 250-300 | – | 50.0 | 21.0 |
| 400 | – | 62.5 | 21.0 |
| 500 | – | 62.5 | 25.0 |
| 600-750 | – | 75.0 | 25.0 |
| 800-1000 | – | 83.25 | 33.0 |
| 1250-2000 | – | 83.26 | 42.0 |

| SCREW TORQUES | |
|---|---|
| Screw Size, Inches Across, Hex Flats | Torque, Pound-Feet |
| 1/8 | 4.2 |
| 5/32 | 8.3 |
| 3/16 | 15 |
| 7/32 | 23.25 |
| 1/4 | 42 |

## CONSTRUCTION SAFETY AND HEALTH RESOURCES

Most resource materials can be found on the OSHA Web site (www.osha.gov).

### Publications
Publications can be downloaded at or ordered from www.osha.gov/pls/publications/pubindex.list

*A Guide to Scaffold Use in the Construction Industry*
(booklet, in question-and-answer format, highlights information about scaffold safety)
OSHA Publication 3150 (Revised 2002), 2.1 MB PDF, 73 pages.
www.osha.gov/Publications/osha3150.pdf

*Concrete and Masonry Construction*
(details about OSHA's concrete and masonry standard)
OSHA Publication 3106 (Revised 1998), 414 KB PDF, 32 pages.
www.osha.gov/Publications/osha3106.pdf

*Crystalline Silica Exposure Card for Construction*
(discusses silica hazards and what employers and employees can do to protect against exposures to silica)
OSHA Publication 3177 (Revised 2002), 2 pages.
A Spanish version is also available. OSHA Publication 3179 (Revised 2003), 2 pages
www.osha.gov/Publications/osha3177.pdf

*Excavations*
(explains all aspects of excavation and trenching)
OSHA Publication 2226 (Revised 2002), 533 KB PDF, 44 pages.
www.osha.gov/Publications/osha2226.pdf

*Fall Protection in Construction*
OSHA Publication 3146 (Revised 1998), 177 KB PDF, 43 pages.
www.osha.gov/Publications/osha3146.pdf

## CONSTRUCTION SAFETY AND
## HEALTH RESOURCES *(cont.)*

*Ground-Fault Protection on Construction Sites*
(booklet about ground-fault circuit interrupters for safe use of
portable tools)
OSHA Publication 3007 (Revised 1998), 100 KB PDF, 31 pages.
www.osha.gov/Publications/osha3007.pdf

*Lead in Construction*
(describes hazards and safe work practices concerning lead)
OSHA Publication 3142 (Revised 2003), 610 KB PDF, 38 pages.
www.osha.gov/Publications/osha3142.pdf

*OSHA Assistance for the Residential Construction Industry*
(many OSHA standards apply to residential construction for
the prevention of possible fatalities. This Web page provides
information about those standards and the hazards present in
residential construction. It was developed in cooperation with
the National Association of Home Builders (NAHB) as part of
the OSHA-NAHB Alliance)
www.osha.gov/SLTC/residential/index.html

*Selected Construction Regulations (SCOR) for the Home
Building Industry*
(provides information on safe and healthful work practices for
residential construction employers; identifies OSHA standards
applicable to hazards found at worksites in the residential
construction industry)
OSHA Publication (Revised 1997), 1.2 MB PDF, 224 pages.
www.osha.gov/Publications/scor1926.pdf

*Stairways and Ladders*
(explains OSHA requirements for stairways and ladders)
OSHA Publication 3124 (Revised 2003), 155 KB PDF, 15 pages.
www.osha.gov/Publications/osha3124.pdf

## CONSTRUCTION SAFETY AND
## HEALTH RESOURCES *(cont.)*

*Working Safely in Trenches*
(provides safety tips for workers in trenches; Spanish version
is on the reverse side)
OSHA Publication 3243 (2005), 2 pages.
www.osha.gov/Publications/trench/trench_safety_tips_card.pdf

### Crane Safety
*Safety and Health Topics: Crane, Derrick and Hoist Safety—
Hazards and Possible Solutions*
(OSHA Web site index provides references to aid in identifying
crane, derrick and hoist hazards in the workplace)
December 2003. One page.
www.osha.gov/SLTC/cranehoistsafety/recognition.html

### Electrical Hazards
*Control of Hazardous Energy (Lockout/Tagout)*
(presents OSHA's general requirements for controlling
hazardous energy during service or maintenance of machines
or equipment)
OSHA Publication 3120 (Revised 2002), 174 KB PDF, 45 pages.
www.osha.gov/Publications/osha3120.pdf

*Controlling Electrical Hazards*
(provides an overview of basic electrical safety on the job)
OSHA Publication 3075 (Revised 2002), 349 KB PDF, 71 pages.
www.osha.gov/Publications/osha3075.pdf

*Safety and Health Topics: Lockout/Tagout*
(OSHA Web site index to information about lockout/tagout,
including hazard recognition, compliance, standards and
directives, Review Commission and Administrative Law Judge
Decisions, standard interpretations and compliance letters,
compliance assistance and training)
www.osha.gov/SLTC/controlhazardousenergy/index.html

## CONSTRUCTION SAFETY AND HEALTH RESOURCES *(cont.)*

### Hazard Communication

*Hazard Communication: Foundation of Workplace Chemical Safety Programs*
(OSHA Web site index for resources on hazard communication)
www.osha.gov/SLTC/hazardcommunications/index.html

*Frequently Asked Questions for Hazard Communication*
(Web site questions and answers about hazard communication)
OSHA, 6 pages.
www.osha.gov/html/faq-hazcom.html

*Hazard Communication Standard*
(highlights protections under OSHA's Hazard Communication standard)
OSHA Fact Sheet No. 93-26 (1993), 3 pages.
www.osha.gov/pls/oshaweb/owadisp.show_document?
p_table=FACT_SHE_ETS&p_id=151

*Hazard Communication Guidelines for Compliance*
(aids employers in understanding the Hazard Communication standard and in implementing a hazard communication program)
OSHA Publication 3111 (2000), 112 KB PDF, 33 pages.
www.osha.gov/Publications/osha3111.pdf

*Chemical Hazard Communication*
(this booklet answers several basic questions about chemical hazard communication)
OSHA Publication 3084 (1998), 248 KB PDF, 31 pages.
www.osha.gov/Publications/osha3084.pdf

*NIOSH Pocket Guide to Chemical Hazards*
(handy source of general industrial hygiene information on several hundred chemicals/classes for workers, employers and occupational health professionals)
www.cdc.gov/niosh/npg/npg.html

# CONSTRUCTION SAFETY AND
# HEALTH RESOURCES *(cont.)*

## Material Handling
*Materials Handling and Storage*
(a comprehensive guide to hazards and safe work practices in
handling materials)
OSHA Publication 2236 (Revised 2002), 559 KB PDF, 40 pages.
www.osha.gov/Publications/osha2236.pdf

## Personal Protective Equipment
*Personal Protective Equipment*
(discusses equipment most commonly used for protection for
the head, including eyes and face and the torso, arms, hands
and feet, the use of equipment to protect against life-threatening hazards is also discussed)
OSHA Publication 3155 (2003), 305 KB PDF, 44 pages.
www.osha.gov/Publications/osha3151.pdf

*Safety and Health Topics: Personal Protective Equipment*
(OSHA Web site index to hazard recognition, control and
training related to personal protective equipment)
www.osha.gov/SLTC/personalprotectiveequipment/index.html

## Toxic Metals: Cadmium
*Safety and Health Topics: Cadmium*
(OSHA Web site index to recognition, evaluation, control,
compliance and training related to Cadmium)
www.osha.gov/SLTC/cadmium/index.html

## Electronic Construction Resources
OSHA eTools and Expert Advisors can be found on OSHA's
Web site: www.osha.gov

# CHAPTER 2
## *Scaffolding and Ladders*

| SCAFFOLD LOADS | |
|---|---|
| **Rated Load Capacity** | **Intended Load** |
| Light-duty | 25 pounds per square foot applied uniformly over the entire span area. |
| Medium-duty | 50 pounds per square foot applied uniformly over the entire span area. |
| Heavy-duty | 75 pounds per square foot applied uniformly over the entire span area. |
| One-person | 250 pounds placed at the center of the span (total 250 pounds). |
| Two-person | 250 pounds placed 18" to the left and right of the center of the span (total 500 pounds). |
| Three-person | 250 pounds placed at the center of the span and 250 pounds placed 18" to the left and right of the center of the span (total 750 pounds). |

**Note:** Platform units used to make scaffold platforms intended for light-duty use shall be capable of supporting at least 25 pounds per square foot applied uniformly over the entire unit-span area, or a 250-pound point load placed on the unit at the center of the span, whichever load produces the greater shear force.

## ALLOWABLE SCAFFOLD SPANS

**2" × 10" (nominal) or 2" × 9" (rough) solid sawn wood planks**

| Maximum intended nominal load (pounds per square foot) | Maximum permissible span using full thickness undressed lumber (feet) | Maximum permissible span using nominal thickness lumber (feet) |
|---|---|---|
| 25 | 10 | 8 |
| 50 | 8 | 6 |
| 75 | 6 | — |

## MAXIMUM NUMBER OF PLANKED LEVELS

| Number of Working Levels | Maximum number of additional planked levels | | | Maximum height of scaffold (feet) |
| | Light duty | Medium duty | Heavy duty | |
|---|---|---|---|---|
| 1 | 16 | 11 | 6 | 125 |
| 2 | 11 | 1 | 0 | 125 |
| 3 | 6 | 0 | 0 | 125 |
| 4 | 1 | 0 | 0 | 125 |

## SAFETY NETS

Outward from the outermost projection of the work surface

| Vertical distance from working level to horizontal plane of net | Minimum required horizontal distance of outer edge of net from the edge of the working surface |
|---|---|
| Up to 5' | 8' |
| More than 5', up to 10' | 10' |
| More than 10' | 13' |

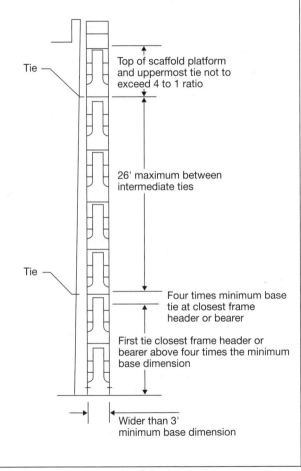

# SCAFFOLD MAXIMUM VERTICAL TIE SPACING WIDER THAN 3' BASES

Tie

Top of scaffold platform and uppermost tie not to exceed 4 to 1 ratio

26' maximum between intermediate ties

Tie

Four times minimum base tie at closest frame header or bearer

First tie closest frame header or bearer above four times the minimum base dimension

Wider than 3' minimum base dimension

2-3

# SYSTEM SCAFFOLD

Joint connections vary according to manufacturer

Guard rail system

Toeboard

Working level

Posts

Runners

Stair tower

Bearers

Screw jack

Sills

Diagonal braces

# TUBE AND COUPLER SCAFFOLD

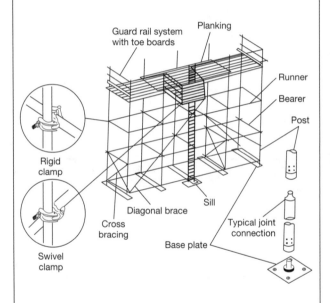

Guard rail system with toe boards

Planking

Runner

Bearer

Post

Rigid clamp

Swivel clamp

Diagonal brace

Cross bracing

Sill

Typical joint connection

Base plate

All ties should be located at clamp locations.

## SINGLE WOOD POLE SCAFFOLDS

| | Light duty up to 20' high | Light duty up to 60' high | Medium duty up to 60' high | Heavy duty up to 60' high |
|---|---|---|---|---|
| Maximum intended load | 25 lb/ft$^2$ | 25 lb/ft$^2$ | 50 lb/ft$^2$ | 75 lb/ft$^2$ |
| Poles or uprights | 2" × 4" | 4" × 4" | 4" × 4" | 4" × 6" |
| Maximum longitudinal pole spacing | 6' | 10' | 8' | 6' |
| Maximum transverse pole spacing | 5' | 5' | 5' | 5' |
| Runners | 1" × 4" | 1¼" × 9" | 2" × 10" | 2" × 10" |
| Bearers and maximum spacing of bearers | | | | |
| 3' | 2" × 4" | 2" × 4" | 2" × 10" or 3" × 4" | 2" × 10" or 3" × 5" |
| 5' | 2" × 6" or 3" × 4" | 2" × 6" or 3" × 4" (rough) | 2" ×10" or 3" × 4" | 2" × 10" or 3" × 5" |
| 6' | — | — | 2" × 10" or 3" × 4" | 2" × 10" or 3" × 5" |
| 8' | — | — | 2" × 10" or 3" × 4" | — |

## SINGLE WOOD POLE SCAFFOLDS *(cont.)*

| | Light duty up to 20' high | Light duty up to 60' high | Medium duty up to 60' high | Heavy duty up to 60' high |
|---|---|---|---|---|
| Planking | $1\frac{1}{4}$" × 9" | 2" × 10" | 2" × 10" | 2" × 10" |
| Maximum vertical spacing of horizontal members | 7' | 9' | 7' | 6' 6" |
| Horizontal bracing | 1" × 4" | 1" × 4" | 1" × 6" or $1\frac{1}{4}$" × 4" | 2" × 4" |
| Diagonal bracing | 1" × 4" | 1" × 4" | 1" × 4" | 2" × 4" |
| Tie-ins | 1" × 4" | 1" × 4" | 1" × 4" | 1" × 4" |

**Note:** All members except planking are used on edge. All wood bearers shall be reinforced with a $\frac{3}{16}$" × 2" steel strip, or the equivalent, secured to the lower edges for the entire length of the bearer.

| INDEPENDENT WOOD POLE SCAFFOLDS | | | | |
|---|---|---|---|---|
| | Light duty up to 20' high | Light duty up to 60' high | Medium duty up to 60' high | Heavy duty up to 60' high |
| Maximum intended load | 25 lb/ft² | 25 lb/ft² | 50 lb/ft² | 75 lb/ft² |
| Poles or uprights | 2" × 4" | 4" × 4" | 4" × 4" | 4" × 4" |
| Maximum longitudinal pole spacing | 6' | 10' | 8' | 6' |
| Maximum transverse pole spacing | 6' | 10' | 8' | 8' |
| Runners | 1¼" × 4" | 1¼" × 9" | 2" × 10" | 2" × 10" |
| Bearers and maximum spacing of bearers | | | | |
| 3' | 2" × 4" | 2" × 4" | 2" × 10" | 2" × 10" (rough) |
| 6' | 2" × 6" or 3" × 4" | 2" × 10" (rough) or 3" × 8" | 2" × 10" | 2" × 10" (rough) |
| 8' | 2" × 6" or 3" × 4" | 2" × 10" (rough) or 3" × 8" | 2" × 10" | — |
| 10' | 2" × 6" or 3" × 4" | 2" × 10" (rough) or 3" × 8" | — | — |

## INDEPENDENT WOOD POLE SCAFFOLDS *(cont.)*

| | Light duty up to 20' high | Light duty up to 60' high | Medium duty up to 60' high | Heavy duty up to 60' high |
|---|---|---|---|---|
| Planking | 1¼" × 9" | 2" × 10" | 2" × 10" | 2" × 10" |
| Maximum vertical spacing of horizontal members | 7' | 7' | 6' | 6' |
| Horizontal bracing | 1" × 4" | 1" × 4" | 1" × 6" or 1¼" × 4" | 2" × 4" |
| Diagonal bracing | 1" × 4" | 1" × 4" | 1" × 4" | 2" × 4" |
| Tie-ins | 1" × 4" | 1" × 4" | 1" × 4" | 1" × 4" |

**Note:** All members except planking are used on edge. All wood bearers shall be reinforced with a ³⁄₁₆" × 2" steel strip, or the equivalent, secured to the lower edges for the entire length of the bearer.

## MINIMUM SIZE OF MEMBERS — TUBE SCAFFOLDS

| | Light duty | Medium duty | Heavy duty |
|---|---|---|---|
| Maximum intended load | 25 lb/ft$^2$ | 50 lb/ft$^2$ | 75 lb/ft$^2$ |
| Posts, runners and braces | Nominal 2" (1.90") OD steel tube or pipe | Nominal 2" (1.90") OD steel tube or pipe | Nominal 2" (1.90") OD steel tube or pipe |
| Bearers | Nominal 2" (1.90")<br><br>OD steel tube or pipe and a maximum post spacing of 4' × 10' | Nominal 2" (1.90")<br><br>OD steel tube or pipe and a maximum post spacing of 4' × 7' or nominal 2½" (2.375") OD steel tube or pipe and a maximum post spacing of 6' × 8'* | Nominal 2½" (2.375")<br><br>OD steel tube or pipe and a maximum post spacing of 6' × 6' |
| Maximum vertical runner spacing | 6' 6" | 6' 6" | 6' 6" |

*Bearers shall be installed in the direction of the shorter dimension.

**Note:** Longitudinal diagonal bracing shall be installed at an angle of 45 degrees (+/− 5 degrees).

## SCHEDULE FOR LADDER-TYPE PLATFORMS

| Length of Platform | 12' | 14' and 16' | 18' and 20' | 22' and 24' | 28' and 30' |
|---|---|---|---|---|---|
| **Side stringers, minimum cross section (finished sizes):** | | | | | |
| At ends | $1^{3}/_{4}$" × $2^{3}/_{4}$" | $1^{3}/_{4}$" × $2^{3}/_{4}$" | $1^{3}/_{4}$" × 3" | $1^{3}/_{4}$" × 3" | $1^{3}/_{4}$" × $3^{1}/_{2}$" |
| At middle | $1^{3}/_{4}$" × $3^{3}/_{4}$" | $1^{3}/_{4}$" × $3^{3}/_{4}$" | $1^{3}/_{4}$" × 4" | $1^{3}/_{4}$" × $4^{1}/_{4}$" | $1^{3}/_{4}$" × 5" |
| Reinforcing strip (minimum) | A $1/_{8}$" × $7/_{8}$" steel reinforcing strip shall be attached to the side or underside, full length. | | | | |
| Rungs | Rungs shall be $1^{1}/_{8}$" minimum diameter with at least $7/_{8}$" in diameter tenons, and the maximum spacing shall be 12" to center. | | | | |
| Tie rods (minimum number) | 3 | 4 | 4 | 5 | 6 |
| Diameter (minimum) | $1/_{4}$" | $1/_{4}$" | $1/_{4}$" | $1/_{4}$" | $1/_{4}$" |
| Flooring (minimum finished size) | $1/_{2}$" × $2^{3}/_{4}$" | $1/_{2}$" × $2^{3}/_{4}$" | $1/_{2}$" × $2^{3}/_{4}$" | $1/_{2}$" × $2^{3}/_{4}$" | $1/_{2}$" × $2^{3}/_{4}$" |

## OSHA SCAFFOLDING — KEY PROVISIONS

- **Fall protection or fall arrest systems.** Each employee higher than 10' (3.05 m) above a lower level shall be protected from falls by guardrails or a fall arrest system, except those on single-point and two-point adjustable suspension scaffolds. Each employee on a single-point or two-point adjustable suspended scaffold shall be protected by both a personal fall arrest system and a guardrail.

- **Guardrail height.** The height of the toprail for scaffolds manufactured and placed in service after January 1, 2000, must be between 38" (0.9 m) and 45" (1.2 m). The height of the toprail for scaffolds manufactured and placed in service before January 1, 2000, can be between 36" (0.9) and 45" (1.2 m).

- **Crossbracing.** When the crosspoint of crossbracing is used as a toprail, it must be between 38" (0.97 m) and 48" (1.3 m) above the work platform.

- **Midrails.** Midrails must be installed approximately halfway between the toprail and the platform surface. When a crosspoint of crossbracing is used as a midrail, it must be between 20" (0.5 m) and 30" (0.8 m) above the work platform.

- **Footings.** Support scaffold footings shall be level and capable of supporting the loaded scaffold. The legs, poles, frames and uprights shall bear on base plates and mud sills.

## OSHA SCAFFOLDING — KEY PROVISIONS *(cont.)*

- **Platforms.** Supported scaffold platforms shall be fully planked or decked.
- **Guying ties and braces.** Supported scaffolds with a height-to-base ratio of more than 4:1 shall be restrained from tipping by guying, tying, bracing or the equivalent.
- **Capacity.** Scaffolds and scaffold components must support at least four times the maximum intended load. Suspension scaffold rigging must support at least six times the intended load.
- **Training.** Employers must train each employee who works on a scaffold on the hazards and the procedures to control the hazards.
- **Inspections.** Before each work shift and after any occurrence that could affect the structural integrity, a competent person must inspect the scaffold and scaffold components for visible defects.
- **Erecting and dismantling.** When erecting and dismantling supported scaffolds, a competent person must determine the feasibility of providing a safe means of access and fall protection for these operations.

## OSHA SCAFFOLDING — COMPETENT PERSON

Defined as "one who is capable of identifying existing and predictable hazards in the surroundings or working conditions, which are unsanitary, hazardous to employees, and who has authorization to take prompt corrective measures to eliminate them."

The standard requires a competent person to perform the following duties under these circumstances:

- **In general:**
  - To select and direct employees who erect, dismantle, move or alter scaffolds.
  - To determine whether it is safe for employees to work on or from a scaffold during storms or high winds and to ensure that a personal fall arrest system or wind screens protect these employees. (Note: Windscreens should not be used unless the scaffold is secured against the anticipated wind forces imposed.)

- **For training:**
  - To train employees involved in erecting, disassembling, moving, operating, repairing, maintaining or inspecting scaffolds to recognize associated work hazards.

- **For inspections:**
  - To inspect scaffolds and scaffold components for visible defects before each work shift and after any occurrence that could affect the structural integrity and to authorize prompt corrective actions.
  - To inspect ropes on suspended scaffolds prior to each workshift and after every occurrence that could affect the structural integrity and to authorize prompt corrective actions.
  - To inspect manila or plastic (or other synthetic) rope being used for toprails or midrails.

## OSHA SCAFFOLDING — COMPETENT PERSON *(cont.)*

- **For suspension scaffolds:**
  - To evaluate direct connections to support the load.
  - To evaluate the need to secure two-point and multi-point scaffolds to prevent swaying.
- **For erectors and dismantlers:**
  - To determine the feasibility and safety of providing fall protection and access.
  - To train erectors and dismantlers to recognize associated work hazards.
- **For scaffold components:**
  - To determine whether a scaffold will be structurally sound when intermixing components from different manufacturers.
  - To determine whether galvanic action has affected the capacity when using components of dissimilar metals.

## OSHA SCAFFOLDING — OVERSIGHT

### When a qualified person is required

The scaffolding standard defines a qualified person as "one who . . . by possession of a recognized degree, certificate, or professional standing, or who by extensive knowledge, training, and experience—has successfully demonstrated his/her ability to solve or resolve problems related to the subject matter, the work, or the project."

The qualified person must perform the following duties in these circumstances:

- **In general:**
  - To design and load scaffolds in accordance with that design.

- **For training:**
  - To train employees working on the scaffolds to recognize the associated hazards and understand procedures to control or minimize those hazards.

- **For suspension scaffolds:**
  - To design the rigging for single-point adjustable suspension scaffolds.
  - To design platforms on two-point adjustable suspension types that are less than 36" (0.9 m) wide to prevent instability.
  - To make swaged attachments or spliced eyes on wire suspension ropes.

- **For components and design:**
  - To design scaffold components construction in accordance with the design.

### When an engineer is required

The standard requires a registered professional engineer to perform the following duties in these circumstances:

- **For suspension scaffolds:**
  - To design the direct connections of masons' multi-point, adjustable suspension scaffolds.

- **For design:**
  - To design scaffolds that are to be moved when employees are on them.
  - To design pole scaffolds over 60' (18.3 m) in height.
  - To design tube and coupler scaffolds over 125' (38 m) in height.
  - To design fabricated frame scaffolds over 125' (38 m) in height above their base plates.
  - To design brackets on fabricated frame scaffolds used to support cantilevered loads in addition to workers.
  - To design outrigger scaffolds and scaffold components.

## OSHA SCAFFOLDING — MISCELLANEOUS REQUIREMENTS

### Scaffold Access

Employers must provide access when the scaffold platforms are more than 2' (0.6 m) above or below a point of access.

Direct access is acceptable when the scaffold is not more than 14" (36 cm) horizontally and not more than 24" (61 cm) vertically from the other surfaces.

The standard prohibits the use of cross-braces as a means of access. Several types of access are permitted:

- Ladders, (e.g., portable, hook-on, attachable and stairway).
- Stair towers.
- Ramps and walkways.
- Integral prefabricated frames.

### Erecting and Dismantling Supported Scaffolds

Employees erecting and dismantling supported scaffolding must have a safe means of access provided when a competent person has determined the feasibility and analyzed the site conditions.

### Prohibited Types of Scaffolds

Shore and lean-to scaffolds are strictly prohibited.

Also, employees are prohibited from working on scaffolds covered with snow, ice or other slippery materials, except to remove these substances.

### Capacity

Each scaffold and scaffold component must support, without failure, its own weight and at least four times the maximum intended load applied or transmitted to it.

A qualified person must design the scaffolds, which are loaded in accordance with that design.

Scaffolds and scaffold components must not be loaded in excess of their maximum intended loads or rated capacities, whichever is less.

### Capacity *(cont.)*

Load-carrying timber members should be a minimum of 1,500 lb-f/in$^2$ construction grade lumber.

### Platform Construction

Each platform must be planked and decked as fully as possible, with the space between the platform and uprights not wider than 1" (2.5 cm). The space must not exceed 9 inches (24.1 centimeters) when side brackets or odd-shaped structures result in a wider opening between the platform and the uprights.

### Requirements for Planking

Scaffold planking must be able to support, without failure, its own weight and at least four times the intended load.

Solid sawn wood, fabricated planks and fabricated platforms may be used as scaffold planks following the recommendations by the manufacturer, a lumber grading association or an inspection agency.

### Maximum Deflection

The platform must not deflect more than 1/60 of the span when loaded.

### Debris

The standard prohibits work on platforms cluttered with debris.

### Work Area Width

Each scaffold platform and walkway must be at least 18" (46 cm) wide. When the work area is less than 18" (46 cm) wide, guardrails or personal fall arrest systems must be used.

### Guardrails

The standard requires employers to protect each employee on a scaffold more than 10' (3.1 m) above a lower level from falling to that lower level.

To ensure adequate protection, install guardrails along all open sides and ends before releasing the scaffold for use by employees, other than the erection and dismantling crews. Guardrails are *not* required in the following circumstances:

- When the front end of all platforms are less than 14" (36 cm) from the face of the work.
- When outrigger scaffolds are 3" (8 cm) or less from the front edge.
- When employees are plastering and lathing 18" (46 cm) or less from the front edge.

### Materials

Steel or plastic banding must not be used as a toprail or a midrail.

### Supported Scaffolds

Supported scaffolds are platforms supported by legs, outrigger beams, brackets, poles, uprights, posts, frames, or similar rigid support. The structural members, poles, legs, posts, frames, and uprights, must be plumb and braced to prevent swaying and displacement.

## OSHA SCAFFOLDING —
## MISCELLANEOUS REQUIREMENTS *(cont.)*

### Training

All employees must be trained by a qualified person to recognize the hazards associated with the type of scaffold being used and how to control or minimize those hazards. The training must include fall hazards, falling object hazards, electrical hazards, proper use of the scaffold and handling of materials.

### Tipping

Supported scaffolds with a height to base width ratio of more than 4:1 must be restrained by guying, tying, bracing or an equivalent means. Either the manufacturers' recommendation or the following placements must be used for guys, ties and braces:

- Install guys, ties or braces at the closest horizontal member to the 4:1 height and repeat vertically with the top restraint no further than the 4:1 height from the top.
- Vertically—every 20' (6.1 m) or less for scaffolds less than 3' (0.91 m) wide; every 26' (7.9 m) or less for scaffolds more than 3' (0.91 m) wide.
- Horizontally—at each end; at intervals not to exceed 30' (9.1 m) from one end.

### Footing and foundation requirements

Supported scaffolds' poles, legs, posts, frames and uprights must bear on base plates and mud sills, or other adequate firm foundation.

### Forklift support

Forklifts can support platforms only when the entire platform is attached to the fork and the fork-lift does not move horizontally when workers are on the platform.

Front-end loaders and similar equipment can support scaffold platforms only when they have been specifically designed by the manufacturer for such use.

### Materials used to increase the working level height of employees on supported scaffolds

Stilts may be used on a large area scaffold. When a guardrail system is used, the guardrail height must be increased in height equal to the height of the stilts. The manufacturer must approve any alterations to the stilts.

**Note:** A large area scaffold consists of a pole, tube and coupler systems, or a fabricated frame scaffold erected over a substantial portion of the work area.

### SUSPENSION SCAFFOLDS

A suspension scaffold contains one or more platforms suspended by ropes or other nonrigid means from an overhead structure, such as the following scaffolds: single-point, multi-point, multi-level, two-point, adjustable, boatswains' chair, catenary, chimney hoist, continuous run, elevator false car, go-devils, interior hung, masons' and stone setters'.

# SUSPENDED SCAFFOLD WELDING PRECAUTIONS

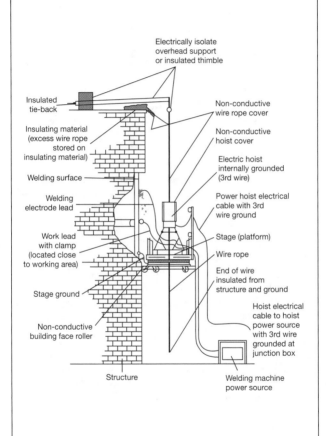

Electrically isolate overhead support or insulated thimble

Insulated tie-back

Insulating material (excess wire rope stored on insulating material)

Welding surface

Welding electrode lead

Work lead with clamp (located close to working area)

Stage ground

Non-conductive building face roller

Non-conductive wire rope cover

Non-conductive hoist cover

Electric hoist internally grounded (3rd wire)

Power hoist electrical cable with 3rd wire ground

Stage (platform)

Wire rope

End of wire insulated from structure and ground

Hoist electrical cable to hoist power source with 3rd wire grounded at junction box

Structure

Welding machine power source

## SUSPENSION SCAFFOLD REQUIREMENTS

The primary requirements for all types of suspension scaffolds are:

- Employers must ensure that all employees are trained to recognize the hazards associated with the type of scaffold being used.

- All support devices must rest on surfaces capable of supporting at least four times the load imposed on them by the scaffold when operating at the rated load of the hoist, or at least one-and-a-half times the load imposed on them by the scaffold at the stall capacity of the hoist, whichever is greater.

- A competent person must evaluate all direct connections prior to use to confirm that the supporting surfaces are able to support the imposed load.

- All suspension scaffolds must be tied or otherwise secured to prevent them from swaying, as determined by a competent person.

- Guardrails, a personal fall arrest system or both must protect each employee more than 10' (3.1 m) above a lower level from falling.

- A competent person must inspect ropes for defects prior to each work shift and after every occurrence that could affect a rope's integrity.

- When scaffold platforms are more than 24" (61 cm) above or below a point of access, ladders, ramps, walkways or similar surfaces must be used.

- When using direct access, the surface must not be more than 24" (61 cm) above or 14" (36 cm) horizontally from the surface.

- When lanyards are connected to horizontal life-lines or structural members on single-point or two-point adjustable scaffolds, the scaffold must have additional independent support lines equal in number and strength to the suspension lines and have automatic locking devices.

- Emergency escape and rescue devices must not be used as working platforms unless designed to function as suspension scaffolds and emergency systems.

### Requirements for Counterweights

Counterweights used to balance adjustable suspension scaffolds must be able to resist at least four times the tipping moment imposed by the scaffold operating at either the rated load of the hoist, or one-and-a-half (minimum) times the tipping moment imposed by the scaffold operating at the stall load of the hoist, whichever is greater.

Only those items specifically designed as counterweights must be used.

Counterweights used for suspended scaffolds must be made of materials that cannot be easily dislocated. Flowable material, such as sand or water, cannot be used.

Counterweights must be secured by mechanical means to the outrigger beams.

Vertical lifelines must not be fastened to counterweights.

### Sand, Masonry or Roofing Felt Counterweights

Such materials cannot be used as counterweights.

## Outrigger Beams

Outrigger beams (thrust-outs) are the structural members of a suspension or outrigger scaffold that provide support. They must be placed perpendicular to their bearing support.

## Tiebacks

Tiebacks must be secured to a structurally sound anchorage on the building or structure. Sound anchorages do not include standpipes, vents, other piping systems or electrical conduit.

A single tieback must be installed perpendicular to the face of the building or structure. Two tiebacks installed at opposing angles are required when a perpendicular tieback cannot be installed.

## Suspension Ropes

The suspension ropes must be long enough to allow the scaffold to be lowered to the level below without the rope passing through the hoist, or the end of the rope must be configured to prevent the end from passing through the hoist.

The standard prohibits using repaired wire.

Drum hoists must contain no less than four wraps of the rope at the lowest point.

Employers must replace wire rope when the following conditions exist: kinks; six randomly broken wires in one rope lay or three broken wires in one strand in one lay; one-third of the original diameter of the outside wires is lost; heat damage; evidence that the secondary brake has engaged the rope; and any other physical damage that impairs the function and strength of the rope.

Suspension ropes supporting adjustable suspension scaffolds must be a diameter large enough to provide sufficient surface area for the functioning of brake and hoist mechanisms.

Suspension ropes must be shielded from heat-producing processes.

### Power-Operated Hoists

Power-operated hoists used to raise or lower a suspended scaffold must be tested and listed by a qualified testing laboratory.

The stall load of any scaffold hoist must not exceed three times its rated load.

The stall load is the load at which the prime mover (motor or engine) of a power-operated hoist stalls or the power to the prime mover is automatically disconnected.

Gasoline power-operated hoists or equipment are not permitted.

Drum hoists must contain no less than four wraps of suspension rope at the lowest point of scaffold travel.

Gears and brakes must be enclosed.

An automatic braking and locking device, in addition to the operating brake, must engage when a hoist makes an instantaneous change in momentum or an accelerated overspeed.

### Manually Operated Hoists

Manually operated hoists used to raise or lower a suspended scaffold must be tested and listed by a qualified testing laboratory.

These hoists require a positive crank force to descend.

### Welding on a Suspension Scaffold

Welding can be done from suspended scaffolds when:

- A grounding conductor is connected from the scaffold to the structure and is at least the size of the welding lead;
- The grounding conductor is not attached in series with the welding process or the work piece;
- An insulating material covers the suspension wire rope and extends at least 4' (1.2 m) above the hoist;
- Insulated protective covers cover the hoist;
- The tail line is guided, retained or both so that it does not become grounded;
- Each suspension rope is attached to an insulated thimble.
- Each suspension rope and any other independent lines are insulated from grounding.

### Increasing the Working Level Height of Employees on Suspended Scaffolds

No materials or devices may be used to increase the working height on a suspension scaffold. This includes ladders, boxes and barrels.

# FALL PROTECTION

Fall protection includes guardrail systems and personal fall arrest systems. Guardrail systems are referenced in more detail later in this chapter. Personal fall arrest systems include harnesses, components of the harness/belt such as Dee-rings and snap hooks, lifelines and anchorage point.

The following describe the most important points concerning fall protection and personal fall arrest systems.

- Vertical or horizontal lifelines may be used.
- Lifelines must be independent of support lines and suspension ropes and not attached to the same anchorage point as the support or suspension ropes.
- When working from an aerial lift, attach the fall arrest system to the boom or basket.
- Employers must provide fall protection for each employee on a scaffold more than 10' (3.1 m) above a lower level as well as naming a competent person to supervise any and all safety measures.
- A competent person must determine the feasibility and safety of providing fall protection for employees erecting or dismantling supported scaffolds and make sure that all safety measures are in place before any work commences.

## FALL PROTECTION REQUIRED FOR SPECIFIC SCAFFOLDS

| Type of Scaffold | Fall Protection Required |
|---|---|
| Aerial lifts | Personal fall arrest system |
| Boatswains' chair | Personal fall arrest system |
| Catenary scaffold | Personal fall arrest system |
| Crawling board (chicken ladder) | Personal fall arrest system or a guardrail system or by a 3/4" (1.9 cm) diameter grabline or equivalent handhold securely fastened beside each crawling board |
| Float scaffold | Personal fall arrest system |
| Ladder jack scaffold | Personal fall arrest system |
| Needle beam scaffold | Personal fall arrest system |
| Self-contained scaffold | Both a personal adjustable scaffold arrest system and a guardrail system |
| Single-point and two-point suspension scaffolds | Both a personal fall arrest system and a guardrail system |
| Supported scaffold | Personal fall arrest system or guardrail system |
| All other scaffolds not specified above | Personal fall arrest system or guardrail systems that meet the required criteria |

## FALL ARREST SYSTEMS

Personal fall arrest systems can be used on scaffolding when there are no guardrail systems.

Use fall arrest systems when working from the following types of scaffolding: boatswains' chair, catenary, float, needle beam, ladder and pump jack.

Use fall arrest systems also when working from the boom/basket of an aerial lift.

Fall arrest and guardrail systems must be used when working on single- and two-point adjustable suspension scaffolds and self-contained adjustable scaffolds that are supported by ropes.

## FALLING OBJECT PROTECTION

The purpose of falling object requirements is to protect employees from falling hand tools, debris and other small objects by installing toeboards, screens, guardrail systems, debris nets, catch platforms, canopy structures or barricades. In addition, employees are required to wear hard hats.

### Aerial Lift Requirements

Vehicle-mounted aerial devices used to elevate employees—such as extensible boom platforms, aerial lifts, articulating boom platforms and vertical towers—are considered "aerial lifts." Some specific requirements include the following:

- Only authorized personnel can operate aerial lifts.
- The manufacturer or equivalent must certify any modification.

## FALLING OBJECT PROTECTION *(cont.)*

- The insulated portion must not be altered to reduce its insulating value.
- Lift controls must be tested daily.
- Controls must be clearly marked.
- Brakes must be set and outriggers used.
- Boom and basket load limits must not be exceeded.
- Employees must wear personal fall arrest systems, with the lanyard attached to the boom or basket.
- No devices to raise the employee above the basket floor can be used.

### Training

All employees who work on a scaffold must be trained by a person qualified to recognize the hazards associated with the type of scaffold used and to understand the procedures to control and minimize those hazards.

A competent person must train all employees who erect, disassemble, move, operate repair, maintain or inspect scaffolds. Training must cover the nature of the hazard and the correct procedures for erecting, disassembling, moving, operating, repairing, inspecting and maintaining the type of scaffold in use.

Other recommended training topics include erection and dismantling, planning, personal protective equipment, access, guys and braces and parts inspection.

## LADDERS — GENERAL REQUIREMENTS

These rules specify when employers must provide stairways and ladders. In general, the standards require the following:

- When there is a break in elevation of 19" (48 cm) or more and no ramp, runway, embankment or personnel hoist is available, employers must provide a stairway or ladder at all worker points of access.

- When there is only one point of access between levels, employers must keep it clear of obstacles to permit free passage by workers. If free passage becomes restricted, employers must provide a second point of access and ensure that workers use it.

- When there are more than two points of access between levels, employers must ensure that at least one point of access remains clear.

In addition, employers must install all stairway and ladder fall protection systems required by these rules and ensure that their worksite meets all requirements of the stairway and ladder rules before employees use stairways or ladders. See 29 CFR 1926.1050–1060 for the details of the standard.

**Note:** The standard does not apply to ladders specifically manufactured for scaffold access and egress, but does apply to job-made and manufactured portable ladders intended for general purpose use. Rules for ladders used on or with scaffolds are addressed in 29 CFR 1926.451 Subpart L.

## RULES FOR ALL LADDERS

The following rules apply to all ladders:

- Maintain ladders free of oil, grease and other slipping hazards.

- Do not load ladders beyond their maximum intended load nor beyond their manufacturer's rated capacity.

- Use ladders only for their designed purpose.

- Use ladders only on stable and level surfaces unless secured to prevent accidental movement.

- Do not use ladders on slippery surfaces unless secured or provided with slip-resistant feet to prevent accidental movement. Do not use slip-resistant feet as a substitute for exercising care when placing, lashing or holding a ladder on slippery surfaces.

- Secure ladders placed in areas such as passageways, doorways or driveways, or where they can be displaced by workplace activities or traffic to prevent accidental movement. Or use a barricade to keep traffic or activity away from the ladder.

- Keep areas clear around the tops and bottoms of ladders.

- Do not move, shift or extend ladders while in use.

## RULES FOR ALL LADDERS *(cont.)*

- Use ladders equipped with nonconductive side rails if the worker or the ladder could contact exposed energized electrical equipment.

- Face the ladder when moving up or down.

- Use at least one hand to grasp the ladder when climbing.

- Do not carry objects or loads that could cause loss of balance and falling.

In addition, the following general requirements apply to all ladders, including ladders built at the job site:

- **Double-cleated ladders** or two or more ladders must be provided when ladders are the only way to enter or exit a work area where 25 or more employees work or when a ladder serves simultaneous two-way traffic.

- Ladder rungs, cleats and steps must be parallel, level and uniformly spaced when the ladder is in position for use.

- Rungs, cleats and steps of **portable and fixed ladders** (except as provided below) must not be spaced less than 10" (25 cm) apart nor more than 14" (36 cm) apart along the ladder's side rails.

- Rungs, cleats and steps of **step-stools** must not be less than 8" (20 cm) apart nor more than 12" (31 cm) apart between center lines of the rungs, cleats and steps.

## RULES FOR ALL LADDERS *(cont.)*

- Rungs, cleats and steps at the base section of **extension trestle ladders** must not be less than 8" (20 cm) nor more than 18" (46 cm) apart between center lines of the rungs, cleats and steps. The rung spacing on the extension section must not be less than 6" (15 cm) nor more than 12" (31 cm).

- Ladders must not be tied or fastened together to create longer sections unless they are specifically designed for such use.

- When splicing side rails, the resulting side rail must be equivalent in strength to a one-piece side rail made of the same material.

- Two or more separate ladders used to reach an elevated work area must be offset with a platform or landing between the ladders, except when portable ladders are used to gain access to fixed ladders.

- Ladder components must be surfaced to prevent snagging of clothing and injury from punctures or lacerations.

- **Wood ladders** must not be coated with any opaque covering except for identification or warning labels, which may be placed only on one face of a side rail.

**Note:** A competent person must inspect ladders for visible defects periodically and after any incident that could affect their safe use.

## RULES FOR JOB — SPECIFIC LADDERS

- Do not use **single-rail** ladders.
- Use **non-self-supporting ladders** at an angle where the horizontal distance from the top support to the foot of the ladder is approximately one-quarter of the working length of the ladder.
- Use **wooden ladders** built at the job site with spliced side rails at an angle where the horizontal distance is one-eighth of the working length of the ladder.

In addition, the top of a non-self-supporting ladder must be placed with two rails supported equally unless it is equipped with a single support attachment.

## RULES FOR STEPLADDERS

- Do not use the top or top rung of a stepladder as a step.
- Do not use cross bracing on the rear section of stepladders for climbing unless the ladders are designed and provided with steps for climbing on both front and rear sections.
- Metal spreader or locking devices must be provided on stepladders to hold the front and back sections in an open position when ladders are being used.

## RULES FOR PORTABLE LADDERS

The minimum clear distance between side rails for all portable ladders must be 11.5" (29 cm).

In addition, the rungs and steps of portable metal ladders must be corrugated, knurled, dimpled, coated with skid-resistant material or treated to minimize slipping.

Non-self-supporting and self-supporting portable ladders must support at least four times the maximum intended load; extra heavy-duty type 1A metal or plastic ladders must sustain 3.3 times the maximum intended load. To determine whether a self-supporting ladder can sustain a certain load, apply the load to the ladder in a downward vertical direction with the ladder placed at a horizontal angle of 75.5 degrees.

When portable ladders are used for access to an upper landing surface, the side rails must extend at least 3' (0.9 m) above the upper landing surface. When such an extension is not possible, the ladder must be secured and a grasping device such as a grab rail must be provided to assist workers in mounting and dismounting the ladder. A ladder extension must not deflect under a load that would cause the ladder to slip off its supports.

## RULES FOR FIXED LADDERS

If the total length of the climb on a fixed ladder equals or exceeds 24' (7.3 m), the ladder must be equipped with ladder safety devices, self-retracting lifelines and rest platforms at intervals not to exceed 150' (45.7 m), or a cage or well and multiple ladder sections with each ladder section not to exceed 50' (15.2 m) in length. These ladder sections must be offset from adjacent sections, and landing platforms must be provided at maximum intervals of 50' (15.2 m). In addition, fixed ladders must meet the following requirements:

- Fixed ladders must be able to support at least two loads of 250 pounds (114 kg) each, concentrated between any two consecutive attachments. Fixed ladders also must support added anticipated loads caused by ice buildup, winds, rigging and impact loads resulting from using ladder safety devices.

- Individual rung ladders/stepladders must extend at least 42" (1.1 m) above an access level or landing platform either by the continuation of the rung spacings as horizontal grab bars or by providing vertical grab bars that must have the same lateral spacing as the vertical legs of the ladder rails.

- Each step or rung of a fixed ladder must be able to support a load of at least 250 pounds (114 kg) applied in the middle of the step or rung.

# RULES FOR FIXED LADDERS *(cont.)*

- Minimum clear distance between the sides of individual rung ladders/stepladders and between the side rails of other fixed ladders must be 16" (41 cm).

- Rungs of individual rung ladders/stepladders must be shaped to prevent slipping off the end of the rungs.

- Rungs and steps of fixed metal ladders manufactured after March 15, 1991, must be corrugated, knurled, dimpled, coated with skid-resistant material or treated to minimize slipping.

- Minimum perpendicular clearance between fixed ladder rungs, cleats and steps and any obstruction behind the ladder must be 7" (18 cm), except that the clearance for an elevator pit ladder must be 4.5" (11 cm).

- Minimum perpendicular clearance between the centerline of fixed ladder rungs, cleats and steps and any obstruction on the climbing side of the ladder must be 30" (76 cm). If obstructions are unavoidable, clearance may be reduced to 24" (61 cm), provided a deflection device is installed to guide workers around the obstruction.

- Step-across distance between the center of the steps or rungs of fixed ladders and the nearest edge of a landing area must be no less than 7" (18 cm) and no more than 12" (30 cm). A landing platform must be provided if the step-across distance exceeds 12" (30 cm).

## RULES FOR FIXED LADDERS *(cont.)*

- Fixed ladders without cages or wells must have at least a 15" (38 cm) clearance width to the nearest permanent object on each side of the centerline of the ladder.

- Fixed ladders must be provided with cages, wells, ladder safety devices or self-retracting lifelines where the length of climb is less than 24' (7.3 m) but the top of the ladder is at a distance greater than 24' (7.3 m) above lower levels.

- Side rails of through or side-step fixed ladders must extend 42" (1.1 m) above the top level or landing platform served by the ladder. Parapet ladders must have an access level at the roof if the parapet is cut to permit passage through it. If the parapet is continuous, the access level is the top of the parapet.

- Steps or rungs for through-fixed-ladder extensions must be omitted from the extension, and the extension of side rails must be flared to provide between 24" (61 cm) and 30" (76 cm) clearance between side rails.

- When safety devices are provided, the maximum clearance distance between side rail extensions must not exceed 36" (91 cm).

- Fixed ladders must be used at a pitch no greater than 90 degrees from the horizontal, measured from the back side of the ladder.

# CAGES FOR FIXED LADDERS

- Horizontal bands must be fastened to the side rails of rail ladders or directly to the structure, building or equipment for individual-rung ladders.

- Vertical bars must be on the inside of the horizontal bands and must be fastened to them.

- Cages must not extend less than 27" (68 cm) or more than 30" (76 cm) from the centerline of the step or rung and must not be less than 27" (68 cm) wide.

- Insides of cages must be clear of projections.

- Horizontal bands must be spaced at intervals not more than 4' (1.2 m) apart, measured from centerline to centerline.

- Vertical bars must be spaced at intervals not more than 9.5" (24 cm), measured centerline to centerline.

- Bottoms of cages must be between 7' (2.1 m) and 8' (2.4 m) above the point of access to the bottom of the ladder. The bottom of the cage must be flared not less than 4" (10 cm) between the bottom horizontal band and the next higher band.

- Tops of cages must be a minimum of 42" (1.1 m) above the top of the platform or the point of access at the top of the ladder. There must be a way to access the platform or other point of access.

## WELLS FOR FIXED LADDERS

- Wells must completely encircle the ladder.

- Wells must be free of projections.

- Inside faces of wells on the climbing side of the ladder must extend between 27" (68 cm) and 30" (76 cm) from the centerline of the step or rung.

- Inside widths of wells must be at least 30" (76 cm).

- Bottoms of wells above the point of access to the bottom of the ladder must be between 7' (2.1 m) and 8' (2.4 m).

## LADDER SAFETY DEVICES FOR FIXED LADDERS

- The connection between the carrier or lifeline and the point of attachment to the body belt or harness must not exceed 9" (23 cm) in length.

- All safety devices must be able to withstand, without failure, a drop test consisting of a 500-pound (226 kg) weight dropping 18" (41 cm).

- All safety devices must permit the worker to ascend or descend without continually having to hold, push or pull any part of the device, leaving both hands free for climbing.

- All safety devices must be activated within 2' (0.61 m) after a fall occurs and limit the descending velocity of an employee to 7 feet/second (2.1 m/sec) or less.

## REQUIREMENTS FOR MOUNTING LADDER SAFETY DEVICES FOR FIXED LADDERS

The requirements for mounting ladder safety devices for fixed ladders are as follows:

- Mountings for rigid carriers must be attached at each end of the carrier, with intermediate mountings spaced along the entire length of the carrier, to provide the necessary strength to stop workers' falls.

- Mountings for flexible carriers must be attached at each end of the carrier. Cable guides for flexible carriers must be installed with a spacing between 25" (7.6 m) and 40' (12.2 m) along the entire length of the carrier to prevent wind damage to the system. Design and installation of mountings and cable guides must not reduce the strength of the ladder.

- Side rails and steps or rungs for side-step fixed ladders must be continuous in extension.

## RULES FOR DEFECTIVE LADDERS

Ladders needing repairs are subject to the following rules:

- Portable ladders with structural defects (e.g., broken or missing rungs, cleats or steps; broken or split rails; corroded components or other faulty or defective components) must immediately be marked defective or tagged with "Do Not Use" or similar language and withdrawn from service until repaired.

- Fixed ladders with structural defects (e.g., broken or missing rungs, cleats or steps; broken or split rails or corroded components) must be withdrawn from service until repaired.

- Defective fixed ladders are considered withdrawn from use when they are tagged with "Do Not Use" or similar language, or marked in a manner that identifies them as defective or blocked, such as with a plywood attachment that spans several rungs.

- Ladder repairs must restore the ladder to a condition meeting its original design criteria before the ladder is returned to use.

# RULES FOR STAIRWAYS

The rules covering stairways and their components generally depend on how and when stairs are used. Specifically, there are rules for stairs used during construction and stairs used temporarily during construction, as well as rules governing stair rails and handrails.

## Stairways Used During Construction

The following requirements apply to all stairways used during construction:

- Stairways that will not be a permanent part of the building under construction must have landings at least 30" deep and 22" wide (76 × 56 cm) at every 12' (3.7 m) or less of vertical rise.

- Stairways must be installed at least 30 degrees, and no more than 50 degrees, from the horizontal.

- Variations in riser height or stair tread depth must not exceed ¼" (0.64 cm) in any stairway system, including any foundation structure used as one or more treads of the stairs.

- Doors and gates opening directly onto a stairway must have a platform that extends at least 20" (51 cm) beyond the swing of the door or gate.

- Metal pan landings and metal pan treads must be secured in place before filling.

- Stairway parts must be free of dangerous projections such as protruding nails.

## RULES FOR STAIRWAYS *(cont.)*

- Slippery conditions on stairways must be corrected.

- Workers must not use spiral stairways that will not be a permanent part of the structure.

### Temporary Stairs

The following requirements apply to stairways used temporarily during constructions, except during construction of the stairway:

- Do not use stairways with metal pan landings and treads if the treads and/or landings have not been filled in with concrete or other materials unless the pans of the stairs and/or landings are temporarily filled in with wood or other materials. All treads and landings must be replaced when worn below the top edge of the pan.

- Do not use skeleton metal frame structures and steps (where treads and/or landings will be installed later) unless the stairs are fitted with secured temporary treads and landings.

**Note:** Temporary treads must be made of wood or other solid material and installed the full width and depth of the stair.

## RULES FOR STAIRWAYS *(cont.)*

**Stair Rails**

The following general requirements apply to all stair rails:

- Stairways with four or more risers or rising more than 30" (76 cm) in height—whichever is less—must be installed along each unprotected side or edge. When the top edge of a stair rail system also serves as a handrail, the height of the top edge must be no more than 37" (94 cm) nor less than 36" (91.5 cm) from the upper surface of the stair rail to the surface of the tread.

- Stair rails installed after March 15, 1991, must be not less than 36" (91.5 cm) in height.

- Top edges of stair rail systems used as handrails must not be more than 37" (94 cm) high nor less than 36" (91.5 cm) from the upper surface of the stair rail system to the surface of the tread. (If installed before March 15, 1991, not less than 30" [76 cm].)

- Stair rail systems and handrails must be surfaced to prevent injuries such as punctures or lacerations and to keep clothing from snagging.

- Ends of stair rail systems and handrails must be built to prevent dangerous projections, such as rails protruding beyond the end posts of the system.

- Unprotected sides and edges of stairway landings must have standard 42" (1.1 m) guardrail systems.

- Intermediate vertical members, such as balusters used as guardrails, must not be more than 19" (48 cm) apart.
- Other intermediate structural members, when used, must be installed so that no openings are more than 19" (48 cm) wide.
- Screens or mesh, when used, must extend from the top rail to the stairway step and along the opening between top rail supports.

### Handrails

- Handrails and top rails of the stair rail systems must be able to withstand, without failure, at least 200 pounds (890 n) of weight applied within 2" (5 cm) of the top edge in any downward or outward direction at any point along the top edge.
- Handrails must not be more than 37" (94 cm) high nor less than 30" (76 cm) from the upper surface of the handrail to the surface of the tread.
- Handrails must provide an adequate handhold for employees to grasp to prevent falls.
- Temporary handrails must have a minimum clearance of 3" (8 cm) between the handrail and walls, stair rail systems and other objects.
- Stairways with four or more risers, or that rise more than 30" (76 cm) in height—whichever is less—must have at least one handrail.
- Winding or spiral stairways must have a handrail to prevent use of areas where the tread width is less than 6" (15 cm).

## Midrails

Midrails, screens, mesh, intermediate vertical members or equivalent intermediate structural members must be provided between the top rail and stairway steps to the stair rail system. When midrails are used, they must be located midway between the top of the stair rail system and the stairway steps.

## Training Requirements

Employers must train all employees to recognize hazards related to ladders and stairways, and instruct them to minimize these hazards. For example, employers must ensure that each employee is trained by a competent person in the following areas, as applicable:

- Nature of fall hazards in the work area.
- Correct procedures for erecting, maintaining and disassembling the fall protection systems to be used.
- Proper construction, use, placement and care in handling of all stairways and ladders.
- Maximum intended load-carrying capacities of ladders used.

**Note:** Employers must retrain employees as necessary to maintain their understanding and knowledge on the safe use and construction of ladders and stairs.

# CHAPTER 3
## *Personal Protection*

---

### SAFETY CHECKLISTS — PPE

**Personal Protective Equipment (PPE):**

**Eye and Face Protection**
- Safety glasses or face shields are worn anytime work operations can cause foreign objects to get into the eye, such as during welding, cutting, grinding or nailing or when working with concrete and/or harmful chemicals or when exposed to flying particles.
- Eye and face protectors are selected based on anticipated hazards.
- Safety glasses or face shields are worn when exposed to any electrical hazards including work on energized electrical systems.

**Foot Protection**
- Construction workers should wear work shoes or boots with slip-resistant and puncture-resistant soles.
- Safety-toed footwear is worn to prevent crushed toes when working around heavy equipment or falling objects.

## SAFETY CHECKLISTS — PPE *(cont.)*

**Hand Protection**
- Gloves should fit snugly.
- Workers should wear the right gloves for the job (e.g., heavy-duty rubber gloves for concrete work, welding gloves for welding, insulated gloves and sleeves when exposed to electrical hazards).

**Head Protection**
- Workers shall wear hard hats where there is a potential for objects falling from above, bumps to their heads from fixed objects or accidental head contact with electrical hazards.
- Hard hats are routinely inspected for dents, cracks or deterioration.
- Hard hats are replaced after a heavy blow or electrical shock.
- Hard hats are maintained in good condition.

**Floor and Wall Openings**
- Floor openings (12" [30.48 cm] or more) are guarded by a secured cover, a guardrail or equiv-alent on all sides (except at entrances to stair-ways).
- Toeboards are installed around the edges of permanent floor openings (where persons may pass below the opening).

### Scaffolding

- Scaffolds should be set on sound footing.
- Damaged parts that affect the strength of the scaffold are taken out of service.
- Scaffolds are not altered.
- All scaffolds should be fully planked.
- Scaffolds are not moved horizontally while workers are on them unless they are designed to be mobile and workers have been trained in the proper procedures.
- Employees are not permitted to work on scaffolds when scaffolds are covered with snow, ice or other slippery materials.
- Scaffolds are not erected or moved within 10' (3.05 m) of power lines.
- Employees are not permitted to work on scaffolds in bad weather or high winds unless a competent person has determined that it is safe to do so.
- Ladders, boxes, barrels, buckets or other makeshift platforms are not used to raise work height.
- Extra material is not allowed to build up on scaffold platforms.
- Scaffolds should not be loaded with more weight than they were designed to support.

## SAFETY CHECKLISTS — ELECTRICAL

**Electrical Safety**
- Work on new and existing energized (hot) electrical circuits is prohibited until all power is shut off and grounds are attached.
- An effective lockout/tagout system is in place.
- Frayed, damaged or worn electrical cords or cables are promptly replaced.
- All extension cords have grounding prongs.
- Protect flexible cords and cables from damage. Sharp corners and projections should be avoided.
- Use extension cord sets meant for portable electric tools and appliances that are the three-wire type and designed for hard or extra-hard service. (Look for some of the following letters imprinted on the casing: S, ST, SO, STO.)
- All electrical tools and equipment are maintained in safe condition and checked regularly for defects and taken out of service if a defect is found.
- Do not bypass any protective system or device designed to protect employees from contact with electrical energy.
- Overhead electrical power lines are located and identified.
- Ensure that ladders, scaffolds, equipment or materials never come within 10' (3.05 m) of electrical power lines.
- All electrical tools must be properly grounded unless they are double insulated.
- Multiple plug adapters are prohibited.

# SAFETY CHECKLISTS — MISCELLANEOUS

**Elevated Surfaces**
- Signs are posted, when appropriate, showing the elevated surface load capacity.
- Surfaces elevated more than 48" (121.92 cm) above the floor or ground have standard guardrails.
- All elevated surfaces (beneath which people or machinery could be exposed to falling objects) have standard 4" (10.16 cm) toeboards.
- A permanent means of entry and exit with handrails is provided to elevated storage and work surfaces.
- Material is piled, stacked or racked in a way that prevents it from tipping, falling, collapsing, rolling or spreading.

**Hazard Communication**
- A list of hazardous substances used in the workplace is maintained and readily available at the worksite.
- There is a written hazard communication program addressing material safety data sheets (MSDS), labeling and employee training.
- Each container of a hazardous substance (vats, bottles, storage tanks) is labeled with product identity and a hazard warning(s) (communicating the specific health hazards and physical hazards).
- MSDSs are readily available at all times for each hazardous substance used.
- There is an effective employee training program for hazardous substances.

## SAFETY CHECKLISTS — CRANES

**Crane Safety**
- Cranes and derricks are restricted from operating within 10' (3.05 m) of any electrical power line.
- The upper rotating structure supporting the boom and materials being handled is provided with an electrical ground while working near energized transmitter towers.
- Rated load capacities, operating speed and instructions are posted and visible to the operator.
- Cranes are equipped with a load chart.
- The operator understands and uses the load chart.
- The operator can determine the angle and length of the crane boom at all times.
- Crane machinery and other rigging equipment is inspected daily prior to use to make sure that it is in good condition.
- Accessible areas within the crane's swing radius are barricaded.
- Tag lines are used to prevent dangerous swing or spin of materials when raised or lowered by a crane or derrick.

- Illustrations of hand signals to crane and derrick operators are posted on the job site.
- The signal person uses correct signals for the crane operator to follow.
- Crane outriggers are extended when required.
- Crane platforms and walkways have anti-skid surfaces.
- Broken, worn or damaged wire rope is removed from service.
- Guardrails, hand holds and steps are provided for safe and easy access to and from all areas of the crane.
- Load testing reports/certifications are available.
- Tower crane mast bolts are properly torqued to the manufacturer's specifications.
- Overload limits are tested and correctly set.
- The maximum acceptable load and the last test results are posted on the crane.
- Initial and annual inspections of all hoisting and rigging equipment are performed and reports are maintained.
- Only properly trained and qualified operators are allowed to work with hoisting and rigging equipment.

## SAFETY CHECKLISTS — FORKLIFTS

**Forklifts**
- Forklift truck operators are competent to operate these vehicles safely as demonstrated by their successful completion of training and evaluation.
- No employee under 18 years old is allowed to operate a forklift.
- Forklifts are inspected daily for proper condition of brakes, horns, steering, forks and tires.
- Written approval from the truck manufacturer is obtained for any modification or additions that affect capacity and safe operation of the vehicle.
- Capacity, operation and maintenance instruction plates, tags or decals are changed to indicate any modifications or additions to the vehicle.
- Battery charging is conducted in areas specifically designated for that purpose.
- Material handling equipment is provided for handling batteries, including conveyors, overhead hoists or equivalent devices.

## SAFETY CHECKLISTS — FORKLIFTS *(cont.)*

- Reinstalled batteries are properly positioned and secured in the truck.
- Smoking is prohibited in battery charging areas.
- Precautions are taken to prevent open flames, sparks or electric arcs in battery charging areas.
- Refresher training is provided, and an evaluation is conducted whenever a forklift operator has been observed operating the vehicle in an unsafe manner and when an operator is assigned to drive a different type of truck.
- Load and forks are fully lowered, controls neutralized, power shut off and brakes set when a powered industrial truck is left unattended.
- There is sufficient headroom for the forklift and operator under overhead installations, lights, pipes, sprinkler systems, etc.
- Overhead guards are in place to protect the operator against falling objects.
- Trucks are operated at a safe speed.
- All loads are kept stable, safely arranged and fit within the rated capacity of the truck.
- Unsafe and defective trucks are removed from service.

## SELECTION OF RESPIRATORS

| Hazard | Respirator |
|---|---|
| Oxygen deficiency | Self-contained breathing apparatus. Hose mask with blower. Combination air-line respirator with auxiliary self-contained air supply or an air-storage receiver with alarm. |
| Gas and vapor contaminants immediately dangerous to life and health | Self-contained breathing apparatus. Hose mask with blower. Air-purifying full facepiece respirator (for escape only). Combination air-line respirator with auxiliary self-contained air supply or an air-storage receiver with alarm. |
| Gas and vapor contaminants not immediately dangerous to life and health | Air-line respirator. Hose mask without blower. Air-purifying, half-mask or mouthpiece respirator with chemical cartridge. |
| Particulate contaminants immediately dangerous to life and health | Self-contained breathing apparatus. Hose mask with blower. Air-purifying, full facepiece respirator with appropriate filter. Self-rescue mouthpiece respirator (for escape only). Combination air-line respirator with auxiliary self-contained air supply or an air-storage receiver with alarm. |

*Immediately dangerous to life and health* is a condition that poses an immediate threat of severe exposure to contaminants such as radioactive materials, which are likely to have adverse delayed effects on health.

## SELECTION OF RESPIRATORS *(cont.)*

| Hazard | Respirator |
|--------|-----------|
| Particulate contaminants not immediately dangerous to life and health | Air-purifying, half-mask or mouthpiece respirator with filter pad or cartridge. Air-line respirator. Air-line abrasive-blasting respirator. Hose-mask without blower. |
| Combination gas, vapor and particulate contaminants immediately dangerous to life and health | Self-contained breathing apparatus. Hose mask with blower. Air-purifying, full facepiece respirator with chemical canister and appropriate filter (gas mask with filter). Self-rescue mouthpiece respirator (for escape only). Combination air-line respirator with auxiliary self-contained air-supply or an air-storage receiver with alarm. |
| Combination gas, vapor and particulate contaminants not immediately dangerous to life and health | Air-line respirator. Hose mask without blower. Air-purifying, half-mask or mouthpiece respirator with chemical cartridge and appropriate filter. |

*Immediately dangerous to life and health* is a condition that poses an immediate threat of severe exposure to contaminants such as radioactive materials, which are likely to have adverse delayed effects on health.

# EYE AND FACE PROTECTORS

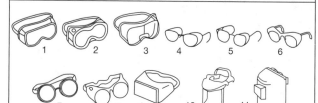

1.  GOGGLES, Flexible Fitting – Regular Ventilation
2.  GOGGLES, Flexible Fitting – Hooded Ventilation
3.  GOGGLES, Cushioned Fitting – Rigid Body
4.  SPECTACLES, Metal Frame – with Sideshields[1]
5.  SPECTACLES, Plastic Frame – with Sideshields
6.  SPECTACLES, Metal-Plastic Frame – with Sideshields[1]
7.  WELDING GOGGLES, Eyecup Type – Tinted Lenses
7A. CHIPPING GOGGLES, Eyecup Type – Clear Safety Lenses
8.  WELDING GOGGLES, Coverspec Type – Tinted Lenses
8A. CHIPPING GOGGLES, Coverspec Type – Clear Safety Lenses
9.  WELDING GOGGLES, Coverspec Type – Tinted Plate Lens
10. FACE SHIELD (Available with Plastic or Mesh Window)
11. WELDING HELMETS

[1]Non-side shield spectacles are available for limited hazard use requiring only frontal protection.

## EYE PROTECTION

| Operation | Hazards | Recommended Protectors |
|---|---|---|
| Acetylene-Burning, Acetylene-Cutting, Acetylene-Welding | Sparks, harmful rays, molten metal, flying particles | 7, 8, 9 |
| Chemical Handling | Splash, acid burns, fumes | 2, 10 (for severe exposure add 10 over 2) |
| Chipping | Flying particles | 1, 3, 4, 5, 6, 7A, 8A |
| Electric (arc) welding | Sparks, intense rays, molten metal | 9, 11, (11 in combination with 4, 5, 6, in tinted lenses advisable) |
| Furnace operations | Glare, heat, molten metal | 7, 8, 9 (for severe exposure add 10) |
| Grinding-Light | Flying particles | 1, 3, 4, 5, 6, 10 |
| Grinding-Heavy | Flying particles | 1, 3, 7A, 8A (for severe exposure add 10) |
| Laboratory | Chemical splash, glass breakage | 2 (10 when in combination with 4, 5, 6) |
| Machining | Flying particles | 1, 3, 4, 5, 6, 10 |
| Molten metals | Heat, glare, sparks, splash | 7, 8, (10 in combination with 4, 5, 6, in tinted lenses) |
| Spot welding | Flying particles, sparks | 1, 3, 4, 5, 6, 10 |

## FILTER LENS SHADE NUMBERS FOR PROTECTION AGAINST RADIANT ENERGY

| Welding Operation | Shade Number |
|---|---|
| Shielded metal-arc welding 1/16", 3/32", 1/8", 5/32" diameter electrodes | 10 |
| Gas-shielded arc welding (nonferrous) 1/16", 3/32", 1/8", 5/32" diameter electrodes | 11 |
| Gas-shielded arc welding (ferrous) 1/16", 3/32", 1/8", 5/32" diameter electrodes | 12 |
| Shielded metal-arc welding 3/16", 7/32", 1/4" diameter electrodes | 12 |
| 5/16", 3/8" diameter electrodes | 14 |
| Atomic hydrogen welding | 10 –14 |
| Carbon-arc welding | 14 |
| Soldering | 2 |
| Torch brazing | 3 or 4 |
| Light cutting, up to 1" | 3 or 4 |
| Medium cutting, 1" to 6" | 4 or 5 |
| Heavy cutting, over 6" | 5 or 6 |
| Gas welding (light), up to 1/8" | 4 or 5 |
| Gas welding (medium), 1/8" to 1/2" | 5 or 6 |
| Gas welding (heavy), over 1/2" | 6 or 8 |

## SOUND AWARENESS AND SAFETY

### Sound Awareness Changes

The typical range of human hearing is 30 hertz–15,000 hertz. Human hearing recognizes an increase of 20 decibels, such as a stereo sound level increase, as being four times as loud at the higher level than it was at the lower level.

| Awareness in Human Hearing | Decibel Change |
|---|---|
| Noticeably Louder | 10 |
| Easily Audible | 5 |
| Faintly Audible | 3 |

## HEARING PROTECTION LEVELS

Because of the Occupational Safety and Health Act of 1970, hearing protection is mandatory if the following time exposures to decibel levels are exceeded because of possible damage to human hearing.

| Decibel Level | Time Exposure Per Day |
|---|---|
| 115 | 15 minutes |
| 110 | 30 minutes |
| 105 | 1 hour |
| 102 | 1½ hours |
| 100 | 2 hours |
| 97 | 3 hours |
| 95 | 4 hours |
| 92 | 6 hours |
| 90 | 8 hours |

# DECIBEL LEVELS OF SOUNDS

The definition of sound intensity is energy (erg) transmitted per 1 second over a square centimeter surface. Sounds are measured in decibels. A decibel (db) change of 1 is the smallest change detected by humans.

| Hearing Intensity | Decibel Level | Examples of Sounds |
|---|---|---|
| Barely | 0 | Dead silence |
| Audible | | Audible hearing threshold |
| | 10 | Room (sound proof) |
| (Very | 20 | Empty auditorium |
| Light) | | Ticking of a stopwatch |
| Audible | 30 | Soft whispering |
| Light | 40 | People talking quietly |
| | | Quiet street noise without autos |
| Medium | 45 | Telephone operator |
| Loud | 50 | Fax machine in office |
| | 60 | Close conversation |
| Loud | 70 | Stereo system |
| | | Computer printer |
| | 80 | Fire truck/ambulance siren |
| | | Cat/dog fight |
| Extremely | 90 | Industrial machinery |
| Loud | | High school marching band |
| Damage | 100 | Heavy duty grinder in a |
| Possible | | machine/welding shop |
| Damaging | 100+ | Begins ear damage |
| | 110 | Diesel engine of a train |
| | 120 | Lightning strike (thunderstorm) |
| | | 60-ton metal forming factory press |
| | 130 | 60" fan in a bus vacuum system |
| | 140 | Commercial/military jet engine |
| Ear Drum | 194 | Space shuttle engines |
| Shattering | 225 | 16" guns on a battleship |

# CHAPTER 4
## *Worker Safety Instructions*

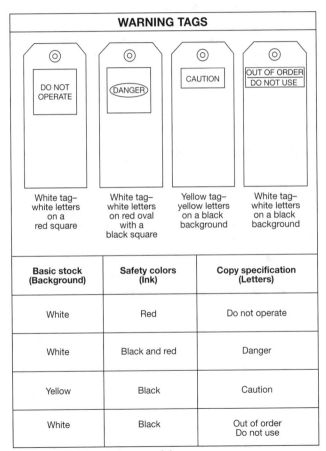

**WARNING TAGS**

|  |  |
|---|---|
| DO NOT OPERATE | White tag–white letters on a red square |
| DANGER | White tag–white letters on red oval with a black square |
| CAUTION | Yellow tag–yellow letters on a black background |
| OUT OF ORDER DO NOT USE | White tag–white letters on a black background |

| Basic stock (Background) | Safety colors (Ink) | Copy specification (Letters) |
|---|---|---|
| White | Red | Do not operate |
| White | Black and red | Danger |
| Yellow | Black | Caution |
| White | Black | Out of order Do not use |

## PERSONAL PROTECTIVE EQUIPMENT (PPE) FOR CONSTRUCTION, DEMOLITION, AND OTHER RELATED WORKPLACES

### Eye and Face Protection

- Safety glasses or face shields are worn anytime work operations can cause foreign objects to get in the eye such as welding, cutting, grinding, nailing or when working with concrete and/or harmful chemicals or when exposed to flying particles. Wear when exposed to any electrical hazards, including working on energized electrical systems.

- Wear eye and face protectors based on the anticipated hazards on the jobsite.

### Foot Protection

- Construction workers should wear work shoes or boots with slip-resistant and puncture-resistant soles.

- Safety-toed footwear is worn to prevent crushed toes when working around heavy equipment or falling objects, especially in demolition.

## PERSONAL PROTECTIVE EQUIPMENT (PPE) FOR CONSTRUCTION, DEMOLITION, AND OTHER RELATED WORKPLACES *(cont.)*

### Hand Protection

- Gloves should fit snugly.

- Workers should wear the right gloves for the job (e.g., heavy-duty rubber gloves for concrete work, welding gloves for welding and insulated gloves and sleeves when exposed to electrical hazards).

### Head Protection

- Wear hard hats where there is a potential for objects falling from above, bumps to the head from fixed objects or of accidental head contact with electrical hazards.

- Routinely inspect hard hats for dents, cracks or deterioration; replace after a heavy blow or electrical shock; maintain in good condition.

### Hearing Protection

- Use earplugs/earmuffs in high-noise work areas where chainsaws, grinders, metal saws and/or heavy equipment are used.

- Clean or replace earplugs regularly.

# RESPIRATORY PROTECTION

Respiratory protection must be worn whenever you are working in a hazardous atmosphere. The appropriate respirator will depend on the contaminant(s) to which you are exposed and the protection factor required. Required respirators must be approved by the National Institute for Occupational Safety and Health (NIOSH), and medical evaluation and training must be provided before use.

### Single-Strap Dust Masks

They are usually not NIOSH-approved and do not protect from hazardous atmospheres. However, they may be useful in providing comfort from pollen or other allergens.

### Approved Dust Masks (Filtering Facepieces)

They can be used for dust, mists, welding fumes, etc. but do not provide protection from gases or vapors. DO NOT USE FOR ASBESTOS OR LEAD. Instead, select from the respirators below.

### Half-Face Respirators

They can be used for protection against most vapors, acid gases, dust or welding fumes. Cartridges/filters must match contaminant(s) and be changed periodically.

## RESPIRATORY PROTECTION (cont.)

**Full-Face Respirators**

They are more protective than half-face respirators. They can also be used for protection against most vapors, acid gases, dust or welding fumes. The face-shield protects face and eyes from Irritants and contaminants. Cartridges/filters must match contaminant(s) and be changed periodically.

**Loose-Fitting Powered-Air-Purifying Respirators**

These offer breathing comfort from a battery-powered fan that pulls air through filters and circulates it throughout the helmet and hood. They can be worn by most workers who have beards. Cartridges/filters must match contaminant(s) and be changed periodically.

**Self-Contained Breathing Apparatus**

This is utilized for entry and escape from atmospheres that are considered immediately dangerous to life and health or are oxygen deficient and are equipped with their own air tank.

## FALL PROTECTION GUIDELINES

- Identify all potential tripping and fall hazards <u>before</u> work starts.
- Look for fall hazards such as unprotected floor openings or floor edges, shafts, stairwells, roof openings or roof edges and skylights.
- Select, wear, and use fall protection equipment appropriate for the task.
- Inspect fall protection equipment for defects before using.
- Stabilize and secure all ladders before climbing them.
- Never stand on the top rung or step of any type of ladder.
- Use handrails when you go up or down stairs.
- Practice good housekeeping. Keep electrical cords and/or cables, welding leads and air hoses out of walkways or adjacent work areas.

## SUPPORTED SCAFFOLD GUIDELINES

Supported scaffolds consist of one or more platforms that are supported by outrigger beams, brackets, poles, legs, uprights, posts, frames or similar rigid support. Guardrails or personal fall arrest systems for fall prevention/protection are required for workers on platforms 10' or higher.

- Working platforms and decks must be planked close to the guardrails.
- Planks are to be overlapped on a support at least 6" but not more than 12".

- Legs, posts, frames, poles and uprights must be on base plates and mud sills, or a firm foundation, and be plumb and braced.

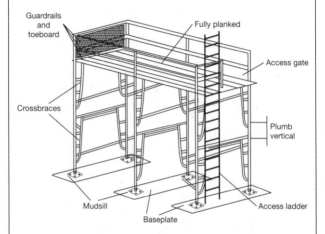

Scaffold user training must include the following:

- The hazards of type of scaffold being used
- Maximum intended load and capacity
- Recognizing and reporting defects
- Fall hazards
- Electrical hazards including overhead lines
- Falling object hazards
- Other hazards that may be encountered

## PORTABLE LADDER SAFETY

Falls from portable ladders (step, straight, combination and extension) are one of the leading causes of occupational fatalities and injuries.

- Read and follow all labels and markings on the ladder.

- Avoid electrical hazards! Look for overhead power lines before handling a ladder. Avoid using a metal ladder near power lines or exposed energized electrical equipment.

- Always inspect the ladder before using it. If the ladder is damaged, it must be removed from service and tagged until repaired or discarded.

- Always maintain a three point (two hands and a foot, or two feet and a hand) contact on the ladder when climbing. Keep your body near the middle of the step and always face the ladder while climbing (see diagram).

- Only use ladders and appropriate accessories (ladder levelers, jacks or hooks) for their designated purposes.

- Ladders must be free of any slippery material on the rungs, steps or feet.

## PORTABLE LADDER SAFETY *(cont.)*

- Do not use a self-supporting ladder (such as a stepladder) as a single ladder or in a partially closed position.

- Do not use the top step/rung of a ladder unless it was designed for that purpose.

- Use a ladder only on a stable and level surface unless it has been secured (top or bottom) to prevent displacement.

- Never place a ladder on a box, barrel or other unstable base to obtain additional height.

- Never move or shift a ladder while a person or equipment is on it.

- An extension or straight ladder used to access an elevated surface must extend at least 3' above the point of support (see diagram). Never stand on the three top rungs of a straight, single or extension ladder.

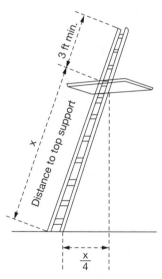

- The proper angle for setting up a ladder is to place its base a quarter of the working length of the ladder from the wall or other vertical surface on previous page.

- A ladder placed in any location where it can be displaced by other work activities must be secured, or a barricade must be erected to keep traffic away from the ladder.
- Be sure that all locks on an extension ladder are properly engaged.
- Never exceed the maximum load rating of a ladder. Always aware of the ladder's load rating and of the weight it is supporting (including the weight of any tools or equipment).

## AERIAL LIFTS

Aerial lifts include boom-supported aerial platforms such as cherry pickers or bucket trucks. The major causes of fatalities are collapses, falls, tip overs and electrocutions.

- Ensure that workers who operate aerial lifts are properly trained in the safe use of the equipment.
- Maintain and operate elevating work platforms in accordance with the manufacturer's instructions.
- Never override hydraulic, mechanical or electrical safety devices.
- Never move the equipment with workers in an elevated platform unless this is clearly stated as permitted by the manufacturer.
- Do not allow workers to position themselves between overhead hazards such as joists and

beams, and the rails of the basket. Any movement of the lift could crush the worker(s).

- Maintain a minimum clearance of at least 10' (3m) from the nearest overhead power lines.
- Always treat power lines, wires and other conductors as energized, even if they are down or appear to be insulated.
- Use a body harness or restraining belt with a lanyard attached to the boom or basket to prevent the worker(s) from being ejected or pulled from the basket.
- Always set the brakes and use wheel chocks when on any type of incline.
- Use outriggers if they are provided.
- Never exceed the load limits of the equipment. Allow for the combined weight of the worker, tools and materials.

## CRANE SAFETY

Serious injuries and fatalities can occur if cranes are not inspected and used properly. Many fatalities can occur when the crane boom, load line or load contacts power lines and shorts electricity to ground. Other incidents happen when workers are struck by the load, are caught inside the swing radius or fail to assemble/disassemble the crane properly.

- Cranes are to be operated only by qualified and trained personnel.
- A designated competent person must inspect the crane and all crane controls before use.

## CRANE SAFETY *(cont.)*

- Always be sure the crane is on a firm and stable surface and is level.

- During assembly or disassembly, never unlock or remove pins unless sections are blocked and secure (stable).

- Fully extend the outriggers and barricade accessible areas inside the crane's swing radius.

- Always look for overhead electric power lines and maintain at least a 10' safe working clearance from these lines.

- Inspect all rigging prior to use and do not wrap hoist lines around the load.

- Use the correct load chart for the crane's current configuration and setup, the weight of the load, and the lift path.

- Do not exceed the load chart capacity.

- Raise load a few inches, hold, verify capacity/balance and test brake system before delivering load.

- Never move loads over workers.

- Be sure to follow proper crane signals and the manufacturer's instructions while operating.

## ELECTRICAL SAFETY

### Safety Tips

- Assume that all overhead wires are energized at lethal voltages. Never assume a wire is safe to touch if it is down or appears insulated.
- Never touch a fallen overhead power line. Call the electric utility company to report fallen electrical lines.
- Stay at least 10' (3 m) away from overhead wires during cleanup and other activities. If working at heights or handling long objects, survey the area before starting work for the presence of overhead wires.
- If an overhead wire falls across your vehicle while you are driving, stay inside the vehicle and continue to drive away from the line. If the engine stalls, do not leave your vehicle. Warn people not to touch the vehicle or the wire. Call or ask someone to call the nearest emergency services and electric utility company.
- Never operate electrical equipment while you are standing in water.
- Never repair electrical cords or equipment unless authorized and qualified.
- Have a qualified electrician inspect electrical equipment that has gotten wet before energizing it.
- If working in damp locations, inspect electric cords and equipment to ensure that they are in good condition and free of defects, and use a ground-fault circuit interrupter.
- Always use caution when working near electricity.

## PORTABLE GENERATOR SAFETY

Portable generators are internal combustion engines used to generate electricity and are commonly used during disaster response. Portable generators can be dangerous if used incorrectly.

### Major Causes of Injuries and Fatalities

- Shocks and electrocution from improper use of power or accidentally energizing other electrical systems.
- Carbon monoxide from a generator's exhaust.
- Fires from improperly refueling the generator or inappropriately storing fuel.

### Safe Work Practices

- Inspect portable generators for damage or loose fuel lines that may have occurred during transportation or handling.
- Keep the generator dry.
- Maintain and operate portable generators in accordance with the manufacturer's use and safety instructions.
- Never attach a generator directly to the electrical system of a structure (home, office or trailer) unless the generator has a properly installed transfer switch, because this creates a risk of electrocution for utility workers.
- Always plug electrical appliances directly into the generator using the manufacturer's supplied cords. Use undamaged heavy-duty extension cords that are grounded (three-pronged).

## PORTABLE GENERATOR SAFETY *(cont.)*

- Use ground-fault circuit interrupters as per the manufacturer's instructions.
- Before refueling, shut down the generator. Never store fuel indoors.

### Carbon Monoxide Poisoning

Carbon monoxide (CO) is a colorless, odorless, toxic gas. Many people have died from CO poisoning because their generator was not adequately ventilated.

- Never use a generator indoors.
- Never place a generator outdoors near doors, windows or vents.
- If you or others show these symptoms—dizziness, headches, nausea, iredness—get to fresh air immediately and seek medical attention.

## TRENCH SAFETY

Never enter an unprotected trench!

Each employee in a trench shall be protected from cave-ins by an adequate protective system.

Some protective systems for trenches are:

- Trench sloped for stability
- Is cut to create stepped benched grades
- Is supported by a system made with posts, beams, shores, or planking and, hydraulic jacks or is supported by a trench box to protect workers

Additionally, excavated or other materials must be at least 2' back from the edge of a trench, and a safe means of egress shall be provided within 25' of workers in a trench.

## DEMOLITION SAFETY

Demolition work involves many of the same hazards that arise during other construction activities. However, demolition involves additional hazards due to a variety of factors. Some of these include lead-based paint, sharp or protruding objects and asbestos-containing material in addition to pipes, pipelines, pumping stations and both above and below ground liquid storage tanks that can contain oil, gasoline, and other extremely hazardous substances. Acids are also very dangerous substances and the proper planning, safety precautions and safety equipment must be utilized before demolition work can start.

- Brace or shore up the walls and floors of structures that have been damaged and that employees must enter.
- Select, wear and use appropriate PPE for the task at hand.
- Inspect all personal protective equipment (PPE) before use.

## DEMOLITION SAFETY *(cont.)*

- Inspect all stairs, passageways, shafts and ladders. Illuminate all stairways and shafts.
- Shut off or cap all electric, gas, water, steam, sewer and other service lines and notify the appropriate utility companies.
- Guard wall openings to a height of 42" cover and secure floor openings with material able to withstand the loads likely to be imposed.
- Floor openings used for material disposal must not be more than 25% of the total floor area.
- Use enclosed chutes with gates on the discharge end to drop demolition material to the ground or into debris containers.
- Demolition of exterior walls and floors must begin at the top of the structure and proceed downward.
- Structural or load-supporting members on any floor must not be cut or removed until all stories above that floor have been removed.
- All roof cornices or other ornamental stonework must be removed prior to pulling walls down.
- Employees must not be permitted to work where structural collapse hazards exist until the hazards are corrected by shoring, bracing or other effective means.

## WORK ZONE TRAFFIC SAFETY

Workers being struck by vehicles or mobile equipment lead to many serious injuries and fatalities. Work zones need traffic controls identified by cones, barrels, barriers, and signs.

Drivers, ground workers and pedestrians must be able to see and understand the proper routes. Construction project managers determine traffic control plans within every type of worksite.

- Traffic control devices such as signals and message boards instruct drivers to follow paths away from active work zones.
- Traffic control devices including cones, barrels, barricades and delineator posts are also used where work is being done.

### Work Zone Protections

Various concrete, water, sand, collapsible barriers, crash cushions and truck-mounted attenuators can help limit motorist intrusions into construction work zones.

### Driving

Rollover protection and seat belts should be used on all moving equipment and vehicles per the manufacturer's recommendations.

## WORK ZONE TRAFFIC SAFETY (cont.)

### Flagging

Flaggers should wear high-visibility clothing with a fluorescent background and made of retroreflective material. This makes workers visible for at least 1,000' in any direction. Check the label or packaging to ensure that the garments are performance class 2 or 3. Drivers should be warned with signs that there will be flaggers ahead. Flaggers should use STOP/SLOW paddles, paddles with lights or flags (only in emergencies).

### Lighting

Flagger stations should be illuminated. Lighting for workers on foot and for equipment operators should be at least 5 foot-candles or greater. Where available lighting is not sufficient, flares or chemical lighting should be used. Glare should be controlled or eliminated.

### Training

Flaggers must be trained/certified and use authorized signaling methods.

## TREE TRIMMING AND REMOVAL SAFETY

**Assume All Power Lines Are Energized!**

- Contact the utility company to discuss de-energizing and grounding or shielding of power lines.
- All tree trimming and removal work within 10' of a power line must be done by trained and experienced line-clearance tree trimmers. A second tree trimmer is required within normal voice communication range.
- Line-clearance tree trimmers must be aware of and maintain the proper minimum approach distances when working around energized power lines.
- Use extreme caution when moving ladders and equipment around downed trees and power lines.

**Stay Alert at All Times!**

- Do not trim trees in dangerous weather conditions.
- Perform a hazard assessment of the work area before starting work.
- Eliminate or minimize exposure to hazards at the tree and in the surrounding area.
- Operators of chain saws and other equipment should be trained and the equipment properly maintained.
- Use personal protective equipment such as gloves, safety glasses, hard hats, hearing protection, etc., as recommended in the equipment manufacturers' operating manuals.

## TREE TRIMMING AND REMOVAL SAFETY *(cont.)*

- Determine the tree's falling direction. Address forward lean, back lean and/or side lean issues.
- Determine the proper amount of hinge wood to safely guide the tree's fall. Provide a retreat path to a safe location.
- Inspect tree limbs for strength and stability before climbing. Tree trimmers working aloft must use appropriate fall protection.
- Do not climb with tools in your hands.
- If broken trees are under pressure, determine the direction of the pressure and make small cuts to release it.
- Use extreme care when felling a tree that has not fallen completely to the ground and is lodged against another tree.
- Never turn your back on a falling tree.
- Be alert! Avoid objects thrown by a falling tree.

# CHAIN SAW SAFETY

Operating a chain saw is inherently hazardous. Potential injuries can be minimized by using the proper personal protective equipment (PPE) along with safe operating procedures.

## Before Starting

- Check controls, chain tension and all bolts and handles to ensure they are functioning properly and are adjusted according to the manufacturer's instructions.
- Make sure the chain is always sharp and the lubrication reservoir is full.
- Start the saw on the ground or on a low firm support. Drop starting is never allowed.
- Start the saw at least 10' from the fueling area, with the chain's brake engaged.

## Fueling

- Use approved containers for transporting fuel to and from the saw.
- Dispense fuel at least 10' away from any sources of ignition when performing construction activities. **No smoking during fueling.**
- Use a funnel or flexible hose when pouring fuel into the saw.
- Never attempt to fuel a hot or running saw.

### Chain Saw Safety Guidelines

- Clear away dirt, debris, small limbs and rocks from the chain path. Check carefully for nails, spikes or other metal in the tree before cutting.
- Shut off the saw or engage its chain brake when carrying on rough or uneven terrain.
- Keep your hands on the saw's handles and maintain secure footing while operating.
- Proper PPE, including hand, foot, leg, eye, face, hearing and head protection must be worn when operating the saw.
- Never wear loose-fitting clothing.
- Be careful that the trunk or tree limbs will not bind against the saw.
- Watch for branches under tension that may spring out when cut.
- Be cautious of saw kick-back. To avoid kick-back, do not saw with the tip. If so equipped, keep the tip guard in place at all times.
- Gasoline-powered chain saws must be equipped with a protective device that minimizes chain saw kickback.

## CHIPPER MACHINE SAFETY

Chipper machines cut tree limbs into small chips. Hazards arise when workers get too close to, or make contact with, the chipper. Contact with chipper operating components (blades, discs or knives) may result in amputation or death. Workers may also be injured by material thrown from the machine. To minimize these hazards, use appropriate engineering, work practice controls, worker training and PPE.

### Chipper Hazards

- Workers making contact with or being pulled into the chipper.
- Hearing loss.
- Face, eye, head or hand injuries from debris.

### Chipper Safety Guidelines

- **NEVER REACH INTO A CHIPPER WHILE IT IS OPERATING**
- Do not wear loose-fitting clothing around a chipper.
- Always follow the manufacturer's guidelines and safety instructions.
- Use earplugs, safety glasses, hard hats and gloves.

## CHIPPER MACHINE SAFETY *(cont.)*

- Workers should be trained on the safe operation of chipper machines. Always supervise new workers using a chipper to ensure that they work safely and never endanger themselves or others.

- Protect yourself from contacting operating chipper components by guarding the infeed and discharge ports and preventing the opening of the access covers or doors until the drum or disc completely stops.

- Prevent detached trailer chippers from rolling or sliding on slopes by chocking the trailer wheels.

- Maintain a safe distance (e.g., two tree or log lengths) between chipper operations and other work/workers.

- When servicing and/or maintaining chipping equipment (e.g., "unjamming"), use a lockout system to ensure that the equipment is de-energized.

## RODENTS, SNAKES, AND INSECTS

### Insects, Spiders and Ticks

- Protect yourself from biting/stinging by wearing long pants, socks, and long-sleeved shirts.
- Use insect repellents that contain DEET or Picaridin.
- Treat bites and stings with over-the-counter products for pain relief and preventing infection.
- Always avoid fire ants because their bites are painful and cause blisters. But there can be severe reactions (chest pain, nausea, sweating, loss of breath, serious swelling or slurred speech) which require immediate medical treatment.

### Rodents and Wild/Stray Animals

- Live and dead animals can spread diseases such as Rat Bite Fever and Rabies.
- Avoid contact with wild or stray animals.
- Avoid contact with rats or rat-contaminated buildings. If you can't avoid contact, wear protective gloves and wash your hands regularly.
- Get rid of dead animals as soon as possible.
- If bitten or scratched, always seek medical attention immediately.

## RODENTS, SNAKES, AND INSECTS *(cont.)*

### Snakes

- Be extremely alert when you are removing debris. If possible, don't place your fingers under debris you are moving. Wear heavy gloves.
- If you see a snake, step back and allow it to proceed at its own pace.
- Wear boots at least 10" high.
- Watch for snakes sunning on fallen trees, limbs, rocks or other debris.
- A snake's striking distance is about half the total length of the snake.
- If bitten, note the color and shape of the snake's head to aid with treatment.
- Keep bite victims still and calm to slow the spread of venom in case the snake is poisonous. Seek medical attention as soon as possible.
- Do not cut the wound or attempt to suck out the venom. Lay the person down so that the bite is below the level of the heat and cover the bite with a clean, dry dressing.

# WEST NILE VIRUS

West Nile Virus infection is an illness transmitted to humans primarily by mosquitoes. Flooded areas, particularly in warm climates, provide ideal conditions for mosquitoes to breed in stagnant water. Bites from infected mosquitoes may result in illnesses that range from mild flu-like conditions (West Nile fever) to severe and sometimes life-threatening diseases requiring hospitalization (West Nile encephalitis or meningitis). If you have symptons of severe illness, seek immediate medical assistance.

## Signs and Symptoms of West Nile Fever (mild illness)

- Headache, fever, body aches.
- Swollen lymph nodes and/or a skin rash on the body.

## Signs and Symptoms of West Nile Encephalitis or Meningitis (severe illness)

- Headache, high fever, stiff neck.
- Disorientation (in very severe cases, coma).
- Tremors, convulsions and muscle weakness (in very severe cases, paralysis).

## Preventing Mosquito Exposure

- Reduce or eliminate mosquito breeding grounds (e.g., sources of stagnant or standing water).
- Cover as much skin as possible by wearing long-sleeved shirts, long pants and socks.
- Avoid use of perfumes and colognes when working outdoors.
- Use an insect repellent containing DEET or Picaridin on uncovered skin areas.
- Choose a repellent that provides protection for the amount of time that you will be exposed. The more DEET or Picaridin a repellent contains, the longer it can protect you.
- Spray insect repellent on the outside of your clothing (mosquitoes can bite through thin clothing).
- Do NOT spray insect repellent on skin that is under clothing.
- Do NOT spray aerosol or pump products in enclosed areas or directly on your face. Do not allow insect repellent to contact your eyes or mouth. Do not use repellents on cuts, wounds or irritated skin.
- After working, use soap and water to wash skin and clothing that has been sprayed with insect repellent.
- Be extremely alert from dusk to dawn, when mosquitoes are most active.

# SILICOSIS

Silicosis is caused by exposure to respirable crystalline silica dust. Crystalline silica is a basic component of soil, sand, granite and most other types of rock, and it is used as an abrasive blasting agent. Silicosis is a progressive, disabling and often fatal lung disease. Cigarette smoking adds to the lung damage caused by silica.

## Effects of Silicosis

- Lung cancer—Silica (has been classified as a human lung carcinogen).
- Bronchitis/Chronic Obstructive Pulmonary Disorder.
- Tuberculosis–Silicosis (makes an individual more susceptible to tuberculosis).
- Scleroderma (a disease affecting skin, blood vessels, joints and skeletal muscles).
- Possible renal disease.

## Symptoms of Silicosis

- Shortness of breath; possible fever.
- Fatigue; loss of appetite.
- Chest pain; dry, nonproductive cough.
- Respiratory failure, which may eventually lead to death.

## SILICOSIS (cont.)

### Sources of Exposure
- Sandblasting for surface preparation.
- Crushing and drilling rock and concrete.
- Masonry and concrete work (e.g., building and road construction and repair).
- Mining/tunneling; demolition work.
- Cement and asphalt pavement manufacturing.

### Preventing Silicosis
- Use all available engineering controls such as blasting cabinets and local exhaust ventilation. Avoid using compressed air for cleaning surfaces.
- Use water sprays and wet methods for cutting, chipping, drilling, sawing, grinding, etc.
- Substitute non-crystalline silica blasting material.
- Use respirators approved for protection against silica; if sandblasting, use abrasive blasting respirators.
- Do not eat, drink or smoke near crystalline silica dust.
- Always wash hands and face before eating, drinking or smoking.

# CARBON MONOXIDE POISONING

Carbon monoxide (CO) is a colorless, odorless, toxic gas that interferes with the oxygen-carrying capacity of blood. CO is non-irritating and can overcome persons without warning. Many people die from CO poisoning, usually while using gasoline-powered tools and generators in buildings or semi-enclosed spaces without adequate ventilation.

## Effects of CO Poisoning

Severe carbon monoxide poisoning causes neurological damage, illness, coma and death.

## Symptoms of CO Exposure

- Headaches, dizziness and drowsiness.
- Nausea, vomiting and tightness across the chest.

## Sources of CO Exposure

- Portable generators/generators in buildings
- Concrete cutting saws, compressors, welding
- Power trowels, floor buffers and space heaters
- Gasoline-powered pumps and vehicles

## CARBON MONOXIDE POISONING (cont.)

### Preventing CO Exposure

- Never use a generator indoors or in enclosed or partially enclosed spaces such as garages, crawl spaces, and basements. Opening windows and doors in an enclosed space may prevent CO buildup.

- Make sure the generator has 3–4' of clear space on all sides and above to ensure adequate ventilation.

- Do not use a generator outdoors if placed near doors, windows or vents that could allow CO to enter and build up in occupied spaces.

- When using space heaters and stoves, ensure that they are in good working order to reduce CO buildup, and never use in enclosed spaces or indoors.

- Consider using tools powered by electricity or compressed air, if available.

- If you experience symptoms of CO poisoning, get to fresh air right away and seek immediate medical attention.

## PERMIT-REQUIRED CONFINED SPACES

A confined space has limited openings for entry or exit, is large enough for entering and working and is not designed for continuous worker occupancy. Confined spaces include underground vaults, tanks, storage bins, manholes, pits, silos, underground utility vaults and pipelines.

**Permit-Required Confined Spaces are Defined as:**

- Containing a hazardous or potentially hazardous atmosphere.
- Containing a material that can engulf an entrant.
- Containing walls that converge inward or floors that slope downward and taper into a smaller area, which could trap or asphyxiate an entrant.
- Containing other serious physical hazards, such as unguarded machines or exposed live wires.
- Having to be identified by the employer, who must inform exposed employees of the existence and location of such spaces and their hazards.

## PERMIT-REQUIRED CONFINED SPACES *(cont.)*

### Safety Guidelines

- Do not enter permit-required confined spaces without being properly trained and without having a permit to enter.
- Review, understand and follow employer's procedures before entering permit-required confined spaces and know how and when to exit them.
- Before entry, identify any physical hazards.
- Before and during entry, test and monitor for oxygen content, flammability, toxicity or explosive hazards as necessary.
- Use employer's fall protection, rescue, air monitoring, ventilation, lighting and communication equipment according to entry procedures.
- Maintain contact at all times with a trained attendant either visually, via phone or by two-way radio. This monitoring system enables the attendant and entry supervisor to order you to evacuate and to alert appropriately trained rescue personnel to safely retrieve entrants when an emergency situation arises.

## GENERAL DECONTAMINATION

Floodwaters may be contaminated with sewage and decaying animal and human remains. Disinfection of hands, clothing, tools/equipment and surfaces in work areas is critical in disease prevention.

### Hand Decontamination

- Wash hands completely with an antibacterial soap and clean water.
- Rinse completely dry with a clean towel or air dry.

### Clothing, Tools and Equipment Decontamination

- It is preferable to use an antibacterial soap and clean water when available.
- If only contaminated water is available, mix $\frac{1}{4}$ cup bleach per gallon of water.
- Immerse objects in solution for 10 minutes; if clothing, gently agitate periodically.
- Transfer objects to hand wash solution for 10 minutes; if clothing, gently agitate periodically.
- Allow clothing, tools and equipment to thoroughly air-dry before re-use.

## Severe Surface Decontamination

Use the following for decontaminating only the most seriously affected surfaces.

- Mix 1½ cups bleach per gallon of water.
- Douse surfaces with the bleach solution and allow to sit for 3 minutes.
- Wipe the contamination from the surface with a paper towel. Then douse the surface again but use the hand wash solution instead.
- Wipe off residual contamination with a paper towel.

## Important Decontamination Considerations

- Always use gloves and eye protection.
- Prepare bleach solutions **daily** and allow to stand for at least 30 minutes before use.
- All containers should be labeled "Bleach-disinfected water, DO NOT DRINK" and "CAUTION: DO NOT MIX BLEACH WITH PRODUCTS CONTAINING AMMONIA."
- Do not immerse electrical or battery-operated tools and equipment in solutions. Clean all exteriors with a rag soaked with an antibacterial soap and clean water or a disinfectant solution.

# LEAD IN CONSTRUCTION/DEMOLITION

Lead is a common hazardous element found at many construction sites. Exposure comes from inhaling fumes and dust, and lead can be ingested when hands are contaminated by lead dust. Lead can be taken home on workers' clothes, skin, hair, tools and in vehicles.

Lead exposure may also take place in demolition sites by way of salvage, removal, encapsulation, renovation and cleanup activities.

## Avoiding Lead Exposure

- Use proper personal protective equipment (e.g., gloves, clothing and approved respirators).
- Wash hands and face before eating and after work.
- Never enter any eating area while wearing protective equipment.
- Never wear clothes and shoes that were worn during lead exposure away from the worksite.
- Launder clothing daily using proper cleaning methods.
- Always alert to symptoms of lead exposure such as severe abdominal pain, headaches and loss of motor coordination.

### Respirator Use

- Wear the appropriate respirators as directed.
- Conduct a user seal check each time a respirator is applied.
- Be aware of your company's respiratory protection program while understanding the limitations and potential hazards of respirators.

### Preventing Further Exposure

- Always ensure adequate ventilation.
- When outdoors, stand upwind of any plume.
- Use dust-collecting equipment when possible.
- Use lead-free chemicals and materials.
- Use wet methods to decrease dust.
- Use local exhaust ventilation for all enclosed work areas.

# HYDROGEN SULFIDE GAS

Hydrogen sulfide ($H_2S$) is a colorless, flammable, extremely hazardous gas with a "rotten egg" smell. It occurs naturally in crude petroleum and natural gas, and can be produced by the breakdown of organic matter and human/animal wastes (e.g., sewage). It is heavier than air and can collect in low-lying and enclosed, poorly ventilated areas such as basements, manholes, sewer lines and underground telephone/electrical vaults.

## Detection Odor

- Can be smelled at low level concentrations, but with continuous low-level exposure or at higher concentrations, you lose your ability to smell the gas even though it is still present.
- At high concentrations your ability to smell the gas can be lost instantly.
- **DO NOT DEPEND ON YOUR SENSE OF SMELL FOR INDICATING THE CONTINUING PRESENCE OF THIS GAS OR FOR ADVANCE WARNING OF HAZARDOUS CONCENTRATIONS.**

## Adverse Health Effects

Health effects vary with how long, and at what concentration level, you are exposed to. Asthmatics may be at greater risk.

## HYDROGEN SULFIDE GAS *(cont.)*

- **Low concentrations**–irritation of eyes, nose, throat or respiratory system; effects can be delayed.
- **Moderate concentrations**–more severe eye and respiratory effects, headache, dizziness, nausea, coughing, vomiting and difficulty breathing.
- **High concentrations**–shock, convulsions, unable to breathe, coma, death; effects can be extremely rapid (within a few breaths).

### Entering Areas with Possible Hydrogen Sulfide

- Before entering, the air must be tested for the presence and concentration of hydrogen sulfide by a qualified person using test equipment. This individual also determines whether fire/explosion precautions are necessary.
- If gas is present, the space should be ventilated.
- If the gas cannot be removed, use appropriate respiratory protection and any other necessary personal protective equipment, rescue, and communication devices. Atmospheres containing high concentrations (greater than 100 ppm) are considered immediately dangerous to life and health, and a self-contained breathing apparatus is required.

# HAND HYGIENE (DISASTER AREAS)

Floodwater can be contaminated with micro-organisms, sewage, industrial waste, chemicals and other substances that can cause illness or death.

## Wear Protective Gloves

- Wear protective gloves when working in contaminated floodwaters, handling contaminated objects or handling human or animal remains.
- Wear thick, cut-resistant gloves made of waterproof material (nitrile or similar washable material).

## Wash Hands with Soap and Clean Water

- Use an antibacterial soap and clean water or a waterless, alcohol-based hand rub.
- Wash hands after cleanup or decontamination work, before preparing or eating food and after toilet use.
- Hand washing prevents disease transmission.

### Wound Care

- Wash wounds with soap and clean water or a waterless, alcohol-based hand rub immediately.
- Seek immediate medical attention if a wound becomes red or swollen or oozes pus.

### Decontamination of Tool/Surface/Equipment

- It is preferable to use soap and clean water.
- If only contaminated water is available, prepare solution of ¼ cup household bleach per 1 gallon of water.
- Prepare fresh solutions daily, preferably just before use.
- Wipe objects with the bleach solution and let stand for 10 minutes; dry thoroughly.
- Label containers (e.g., "bleach disinfected water—DO NOT DRINK").
- **WARNING:** Bleach can damage firefighters' turnout gear; consult manufacturer.

## MOLD EXPOSURE

Molds are microscopic organisms found everywhere in the environment, indoors and outdoors. When present in large quantities, molds have the potential to cause adverse health effects.

### Health Effects of Mold Exposure

- Sneezing
- Cough and congestion
- Runny nose
- Aggravation of asthma
- Eye irritation
- Dermatitis (skin rash)

### People at Greatest Risk

- Individuals with allergies, asthma, sinusitis or other lung diseases.
- Individuals with a weakened immune system (e.g., HIV patients).

### How to Recognize Mold

- Sight — Usually appear as colored woolly mats.
- Smell — Often produce a foul, musty, earthy smell.

### Preventing Mold Growth

- Remove excess moisture with a wet-dry vacuum and dry out the building as quickly as possible.

### Preventing Mold Growth *(cont.)*

- Use fans to assist in the drying process.
- Clean wet materials and surfaces with detergent and water.
- Discard all water-damaged materials.
- Discard all porous materials that have been wet for more than 48 hours.

### General Mold Cleanup Guidelines

- Identify and correct moisture problem.
- Make sure working area is well ventilated.
- Discard mold-damaged materials in plastic bags.
- Clean wet items and surfaces with detergent and water.
- Disinfect cleaned surfaces with ¼ to 1½ cup household bleach in 1 gallon of water.
  **CAUTION: DO NOT MIX BLEACH WITH OTHER CLEANING PRODUCTS THAT CONTAIN AMMONIA.**
- Use respiratory protection. An N-95 respirator is recommended.
- Use hand and eye protection.

## HEAT STRESS GUIDELINES

When the body is unable to cool itself by sweating, several heat-induced illnesses can occur and may result in death.

Factors include high temperature and humidity, direct sun, limited air movement, physical exertion, poor physical condition, certain medications and inadequate tolerance for hot workplaces.

### Preventing Heat Stress
- Know signs/symptoms of heat-related illnesses.
- Block out direct sun or other heat sources.
- Use cooling fans/air-conditioning; rest regularly.
- Drink lots of water; about 1 cup every 15 minutes.
- Wear lightweight, light-colored, loose-fitting clothes.
- Avoid alcohol, caffeinated drinks or heavy meals.

### Heat Exhaustion Symptoms
- Headaches, dizziness, lightheadedness or fainting.
- Weakness, clammy skin, mood changes
- Upset stomach or vomiting.

### Heat Stroke Symptoms
- Dry, hot skin with no sweating.
- Mental confusion or losing consciousness.
- Seizures or fits.

### Heat-Related Illness Guidelines
- Call 911 (or local emergency number) at once.
- Move the worker to a cool, shaded area.
- Loosen or remove heavy clothing.
- Provide cool drinking water.
- Fan and mist the person with water.

# CHAPTER 5
## *Excavation*

### SOIL TYPES

**Type A** soils are cohesive soils with an unconfined, compressive strength of 1.5 tons per square foot (tsf) (144 kPa) or greater. Examples of Type A soils include clay, silty clay, sandy clay, clay loam and, in some cases, silty clay loam and sandy clay loam. Cemented soils such as caliche and hardpan are also considered Type A. However, no soil is Type A if:

- The soil is fissured.
- The soil is subject to vibration from heavy traffic, pile driving or similar effects.
- The soil has been previously disturbed.
- The soil is part of a sloped, layered system where the layers dip into the excavation on a slope of 4 horizontal to 1 vertical (4H:1V) or greater.
- The material is subject to other factors that would require it to be classified as a less stable material.

**Type B** soils are cohesive soils with an unconfined, compressive strength greater than 0.5 tsf (48 kPa) but less than 1.5 tsf (144 kPa). Examples of Type B soils include angular gravel (similar to crushed rock), silt, silt loam, sandy loam and, in some cases, silty clay loam and sandy clay loam; previously disturbed soils except those that would otherwise be classified as Type C soils; soils that meet the unconfined, compressive strength or cementation requirements for Type A, but are fissured or subject to vibration; dry rock that is not stable and material that is part of a sloped, layered system where the layers dip into the excavation on a slope less steep than 4 horizontal to 1 vertical (4H:1V), but only if the material would otherwise be classified as Type B.

**Type C** soils are cohesive soils with an unconfined, compressive strength of 0.5 tsf (48 kPa) or less. Examples of Type C soils include granular soils such as gravel, sand and loamy sand; submerged soil or soil from which water is freely seeping; submerged rock that is not stable and material in a sloped, layered system where the layers dip into the excavation or a slope of 4 horizontal to 1 vertical (4H:1V) or steeper.

**Unconfined compressive strength** refers to the load per unit area at which a soil will fail in compression. It can be determined by laboratory testing or estimated in the field using a pocket penetrometer or by thumb penetration tests and other methods.

**Wet soil** contains significantly more moisture than moist soil, but in such a range of values that cohesive material will slump or begin to flow when vibrated. Granular material that would exhibit cohesive properties when moist will lose those cohesive properties when wet.

## TERMS USED IN GRADING

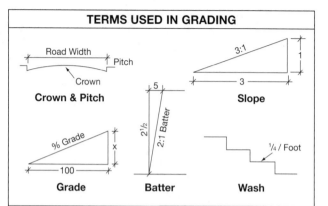

**Crown & Pitch** — Road Width, Pitch, Crown

**Slope** — 3:1, 3, 1

**Grade** — % Grade, x, 100

**Batter** — 5, 2½, 2:1 Batter

**Wash** — ¼ / Foot

## MAXIMUM ALLOWABLE SLOPES

| Soil or Rock Type | Maximum Allowable Slopes (H:V)[1] for Excavations Less Than 20' Deep[3] |
|---|---|
| Stable Rock | Vertical (90 degrees) |
| Type A[2] | ¾:1 (53 degrees) |
| Type B | 1:1 (45 degrees) |
| Type C | 1½:1 (34 degrees) |

[1]Numbers shown in parentheses next to maximum allowable slopes are angles expressed in degrees from the horizontal. Angles have been rounded off.

[2]A short-term maximum allowable slope of ½ H:1V (63 degrees) is allowed in excavations in Type A soil that are 12' (3.67 m) or less in depth. Short-term maximum allowable slopes for excavations greater than 12' (3.67 m) in depth shall be ¾ H:1V (53 degrees).

[3]Sloping or benching for excavations greater than 20' deep should be designed by a registered professional engineer.

# EXCAVATING TOWARD AN OPEN DITCH

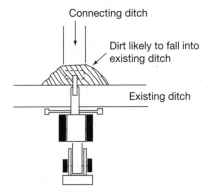

**Hazard From Falling Dirt**

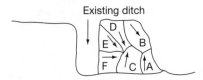

**Excavate in This Sequence to Avoid Falling Dirt**

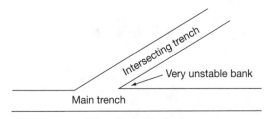

**Very Unstable Bank**

# SEQUENCE OF EXCAVATION IN HIGH GROUNDWATER

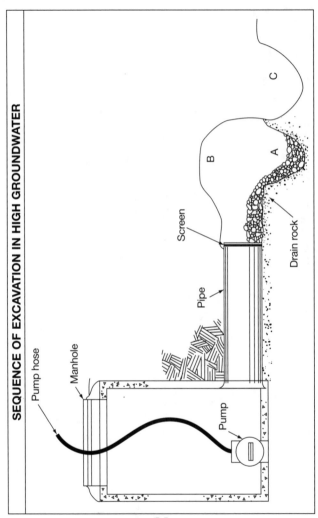

# ANGLE OF REPOSE FOR COMMON SOIL TYPES

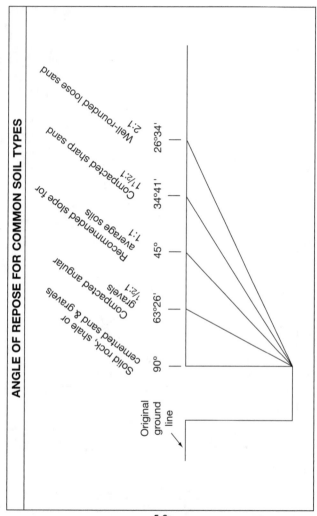

Solid rock, shale or cemented sand & gravels — 90°

Compacted angular gravels — 63°26' — 1/2:1

Recommended slope for average soils — 45° — 1:1

Compacted sharp sand — 34°41' — 1 1/2:1

Well-rounded loose sand — 26°34' — 2:1

Original ground line

5-6

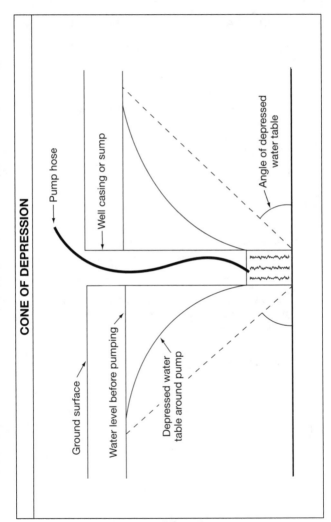

CONE OF DEPRESSION

Pump hose

Well casing or sump

Angle of depressed water table

Ground surface

Water level before pumping

Depressed water table around pump

5-7

## EXCAVATIONS MADE IN TYPE A SOIL

All simple slope excavations 20' or less in depth shall have a maximum allowable slope of ¾:1.

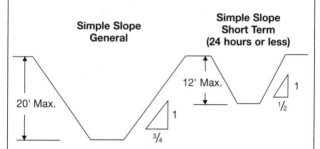

**Simple Slope General**

**Simple Slope Short Term (24 hours or less)**

20' Max.

¾ 1

12' Max.

½ 1

**Exception:** Simple slope excavations that are open 24 hours or less (short term) and that are 12' or less in depth shall have a maximum allowable slope of ½:1.

All benched excavations 20' or less in depth shall have a maximum allowable slope of ¾ to 1 and maximum bench dimensions as follows:

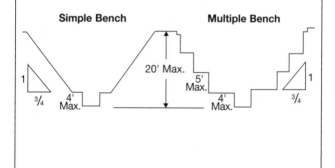

**Simple Bench**

**Multiple Bench**

1 ¾

4' Max.

20' Max.

5' Max.

4' Max.

¾ 1

# EXCAVATIONS MADE IN TYPE B SOIL

All simple slope excavations 20' or less in depth shall have a maximum allowable slope of 1:1.

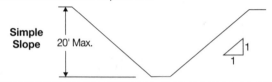

**Simple Slope** — 20' Max.

All benched excavations 20' or less in depth shall have a maximum allowable slope of 1:1 and maximum bench dimensions as follows:

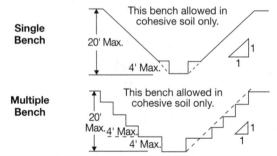

**Single Bench** — 20' Max. — 4' Max.

This bench allowed in cohesive soil only.

**Multiple Bench** — 20' Max. — 4' Max. — 4' Max.

This bench allowed in cohesive soil only.

All excavations 20' or less in depth that have vertically sided lower portions shall be shielded or supported to a height at least 18" above the top of the vertical side. All such excavations shall have a maximum allowable slope of 1:1.

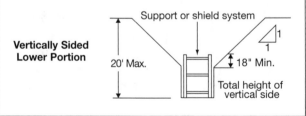

**Vertically Sided Lower Portion** — 20' Max.

Support or shield system

18" Min.

Total height of vertical side

## EXCAVATIONS MADE IN TYPE C SOIL

All simple slope excavations 20' or less in depth shall have a maximum allowable slope of 1½:1.

**Simple Slope**

All excavations 20' or less in depth that have vertically sided lower portions shall be shielded or supported to a height at least 18" above the top of the vertical side. All such excavations shall have a maximum allowable slope of 1½:1.

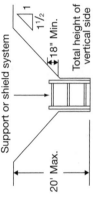

**Vertically Sided Lower Portion**

All excavations 8' or less in depth that have unsupported vertically sided lower portions shall have a maximum vertical side of 3½'.

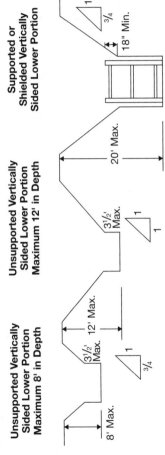

**Unsupported Vertically Sided Lower Portion Maximum 8' in Depth**

**Unsupported Vertically Sided Lower Portion Maximum 12' in Depth**

**Supported or Shielded Vertically Sided Lower Portion**

All excavations more than 8' but not more than 12' in depth with unsupported vertically sided lower portions shall have a maximum allowable slope of 1:1 and a maximum vertical side of 3½'.

All excavations 20' or less in depth that have vertically sided lower portions that are supported or shielded shall have a maximum allowable slope of ¾:1. The support or shield system must extend at least 18" above the top of the vertical side.

# EXCAVATIONS MADE IN LAYERED SOILS

All excavations 20' or less in depth made in layered soils have a maximum allowable slope for each layer.

**B over A**

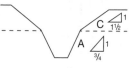

**C over A**

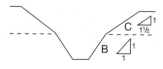

**C over B**

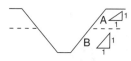

**A over B**

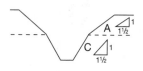

**A over C**

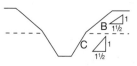

**B over C**

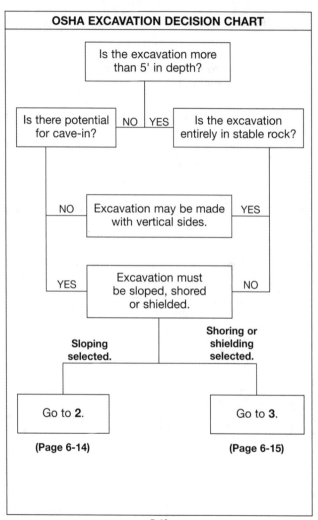

**OSHA EXCAVATION DECISION CHART**

Is the excavation more than 5' in depth?

NO | YES

Is there potential for cave-in?

Is the excavation entirely in stable rock?

NO | Excavation may be made with vertical sides. | YES

YES | Excavation must be sloped, shored or shielded. | NO

**Sloping selected.**

**Shoring or shielding selected.**

Go to **2**.

Go to **3**.

(Page 6-14)

(Page 6-15)

## OSHA EXCAVATION DECISION CHART *(cont.)*

**2**
**Sloping selected as the method of protection.**

Will soil classification be made in accordance with §1926.652 (b)?

YES

NO

Excavation must comply with one of the following three options:

*Option 1:*
§1926.652 (b)(2), which requires Appendices A and B to be followed.

*Option 2:*
§1926.652 (b)(3), which requires other tabulated data to be followed.

*Option 3:*
§1926.652 (b)(4), which requires the excavation to be designed by a registered professional engineer.

Excavations must comply with §1926.652 (b)(1), which requires a slope of 1½H:1V (34°).

## OSHA EXCAVATION DECISION CHART *(cont.)*

> **3**
> **Shoring or shielding**
> **selected as the method**
> **of protection.**

Soil classification is required when
shoring or shielding is used.
The excavation must comply with
one of the following four options:

*Option 1:*
§1926.652 (c)(1), which requires
Appendices A and C to be followed
(e.g., timber shoring).

*Option 2:*
§1926.652 (c)(2), which requires
manufacturers data to be followed
(e.g., hydraulic shoring, trench jacks,
air shores, shields).

*Option 3:*
§1926.652 (c)(3), which requires tabulated
data to be followed (e.g., any system
as per the tabulated data).

*Option 4:*
§1926.652 (c)(4), which requires
the excavation to be designed by a
registered professional engineer
(e.g., any designed system).

# TIMBER TRENCH SHORING MINIMUM TIMBER REQUIREMENTS — TYPE A SOIL

$P(a) = 25 \times H + 72$ psf (2 ft. surcharge)

| Depth of Trench (feet) | Size (actual) and Spacing of Members | | | | | | |
|---|---|---|---|---|---|---|---|
| | Cross Braces | | | | | | |
| | Horiz. Spacing (feet) | Width of Trench (feet) | | | | | Vertical Spacing (feet) |
| | | Up to 4 | Up to 6 | Up to 9 | Up to 12 | Up to 15 | |
| 5 to 10 | Up to 6 | 4 × 4 | 4 × 4 | 4 × 6 | 6 × 6 | 6 × 6 | 4 |
| | Up to 8 | 4 × 4 | 4 × 4 | 4 × 6 | 6 × 6 | 6 × 6 | 4 |
| | Up to 10 | 4 × 6 | 4 × 6 | 4 × 6 | 6 × 6 | 6 × 6 | 4 |
| | Up to 12 | 4 × 6 | 4 × 6 | 6 × 6 | 6 × 6 | 6 × 6 | 4 |
| 10 to 15 | Up to 6 | 4 × 4 | 4 × 4 | 4 × 6 | 6 × 6 | 6 × 6 | 4 |
| | Up to 8 | 4 × 6 | 4 × 6 | 6 × 6 | 6 × 6 | 6 × 6 | 4 |
| | Up to 10 | 6 × 6 | 6 × 6 | 6 × 6 | 6 × 8 | 6 × 8 | 4 |
| | Up to 12 | 6 × 6 | 6 × 6 | 6 × 6 | 6 × 8 | 6 × 8 | 4 |
| 15 to 20 | Up to 6 | 6 × 6 | 6 × 6 | 6 × 6 | 6 × 8 | 6 × 8 | 4 |
| | Up to 8 | 6 × 6 | 6 × 6 | 6 × 6 | 6 × 8 | 6 × 8 | 4 |
| | Up to 10 | 8 × 8 | 8 × 8 | 8 × 8 | 8 × 8 | 8 × 10 | 4 |
| | Up to 12 | 8 × 8 | 8 × 8 | 8 × 8 | 8 × 8 | 8 × 10 | 4 |

# TIMBER TRENCH SHORING MINIMUM TIMBER REQUIREMENTS — TYPE A SOIL (cont.)

$P(a) = 25 \times H + 72$ psf (2 ft. surcharge)

| Depth of Trench (feet) | Size (actual) and Spacing of Members | | | | | | |
|---|---|---|---|---|---|---|---|
| | Wales | | Uprights | | | | |
| | Size (inches) | Vertical Spacing (feet) | Maximum Allowable Horizontal Spacing (feet) | | | | |
| | | | Close | 4 | 5 | 6 | 8 |
| 5 to 10 | Not req'd. | – | – | – | – | 2 × 6 | – |
| | Not req'd. | – | – | – | – | – | 2 × 8 |
| | 8 × 8 | 4 | – | – | 2 × 6 | – | – |
| | 8 × 8 | 4 | – | – | – | 2 × 6 | – |
| 10 to 15 | Not req'd. | – | – | – | – | 3 × 8 | – |
| | 8 × 8 | 4 | – | 2 × 6 | – | – | – |
| | 8 × 10 | 4 | – | – | 2 × 6 | – | – |
| | 10 × 10 | 4 | – | – | – | 3 × 8 | – |
| 15 to 20 | 6 × 8 | 4 | 3 × 6 | – | – | – | – |
| | 8 × 8 | 4 | 3 × 6 | – | – | – | – |
| | 8 × 10 | 4 | 3 × 6 | – | – | – | – |
| | 10 × 10 | 4 | 3 × 6 | – | – | – | – |

Mixed oak or equivalent with a bending strength not less than 850 psi. Manufactured members of equivalent strength may be substituted for wood.

## TIMBER TRENCH SHORING MINIMUM TIMBER REQUIREMENTS — TYPE A SOIL (cont.)

$P(a) = 25 \times H + 72$ psf (2 ft. surcharge)

| Depth of Trench (feet) | Size (S4S) and Spacing of Members | | | | | | |
|---|---|---|---|---|---|---|---|
| | Cross Braces | | | | | | |
| | Horiz. Spacing (feet) | Width of Trench (feet) | | | | | Vertical Spacing (feet) |
| | | Up to 4 | Up to 6 | Up to 9 | Up to 12 | Up to 15 | |
| 5 to 10 | Up to 6 | 4 × 4 | 4 × 4 | 4 × 4 | 4 × 4 | 4 × 6 | 4 |
| | Up to 8 | 4 × 4 | 4 × 4 | 4 × 4 | 4 × 6 | 4 × 6 | 4 |
| | Up to 10 | 4 × 6 | 4 × 6 | 4 × 6 | 6 × 6 | 6 × 6 | 4 |
| | Up to 12 | 4 × 6 | 4 × 6 | 4 × 6 | 6 × 6 | 6 × 6 | 4 |
| 10 to 15 | Up to 6 | 4 × 4 | 4 × 4 | 4 × 4 | 6 × 6 | 6 × 6 | 4 |
| | Up to 8 | 4 × 6 | 4 × 6 | 4 × 6 | 6 × 6 | 6 × 6 | 4 |
| | Up to 10 | 6 × 6 | 6 × 6 | 6 × 6 | 6 × 6 | 6 × 6 | 4 |
| | Up to 12 | 6 × 6 | 6 × 6 | 6 × 6 | 6 × 6 | 6 × 6 | 4 |
| 15 to 20 | Up to 6 | 6 × 6 | 6 × 6 | 6 × 6 | 6 × 6 | 6 × 6 | 4 |
| | Up to 8 | 6 × 6 | 6 × 6 | 6 × 6 | 6 × 6 | 6 × 6 | 4 |
| | Up to 10 | 6 × 6 | 6 × 6 | 6 × 6 | 6 × 6 | 6 × 8 | 4 |
| | Up to 12 | 6 × 6 | 6 × 6 | 6 × 6 | 6 × 8 | 6 × 8 | 4 |

# TIMBER TRENCH SHORING MINIMUM TIMBER REQUIREMENTS — TYPE A SOIL (cont.)

$P(a) = 25 \times H + 72$ psf (2 ft. surcharge)

| Depth of Trench (feet) | Wales | | Uprights | | | | |
|---|---|---|---|---|---|---|---|
| | Size (inches) | Vertical Spacing (feet) | Maximum Allowable Horizontal Spacing (feet) | | | | |
| | | | Close | 4 | 5 | 6 | 8 |
| 5 to 10 | Not req'd. | Not req'd. | – | – | – | 4 × 6 | – |
| | Not req'd. | Not req'd. | – | – | – | – | 4 × 8 |
| | 8 × 8 | 4 | – | – | 4 × 6 | – | – |
| | 8 × 8 | 4 | – | – | – | 4 × 6 | – |
| 10 to 15 | Not req'd. | Not req'd. | – | – | – | 4 × 10 | – |
| | 6 × 8 | 4 | – | 4 × 6 | – | – | – |
| | 8 × 8 | 4 | – | – | 4 × 8 | – | – |
| | 8 × 10 | 4 | – | 4 × 6 | – | 4 × 10 | – |
| 15 to 20 | 6 × 8 | 4 | 3 × 6 | – | – | – | – |
| | 8 × 8 | 4 | 3 × 6 | 4 × 12 | – | – | – |
| | 8 × 10 | 4 | 3 × 6 | – | – | – | – |
| | 8 × 12 | 4 | 3 × 6 | 4 × 12 | – | – | – |

Douglas fir or equivalent with a bending strength not less than 1500 psi. Manufactured members of equivalent strength may be substituted for wood.

# TIMBER TRENCH SHORING MINIMUM TIMBER REQUIREMENTS — TYPE B SOIL

$P(a) = 45 \times H + 72$ psf (2 ft. surcharge)

| Depth of Trench (feet) | Size (actual) and Spacing of Members | | | | | | |
|---|---|---|---|---|---|---|---|
| | Cross Braces | | | | | | |
| | Horiz. Spacing (feet) | Width of Trench (feet) | | | | | Vertical Spacing (feet) |
| | | Up to 4 | Up to 6 | Up to 9 | Up to 12 | Up to 15 | |
| 5 to 10 | Up to 6 | 4 × 6 | 4 × 6 | 6 × 6 | 6 × 6 | 6 × 6 | 5 |
| | Up to 8 | 6 × 6 | 6 × 6 | 6 × 6 | 6 × 8 | 6 × 8 | 5 |
| | Up to 10 | 6 × 6 | 6 × 6 | 6 × 6 | 6 × 8 | 6 × 8 | 5 |
| 10 to 15 | Up to 6 | 6 × 6 | 6 × 6 | 6 × 6 | 6 × 8 | 6 × 8 | 5 |
| | Up to 8 | 6 × 8 | 6 × 8 | 6 × 8 | 8 × 8 | 8 × 8 | 5 |
| | Up to 10 | 8 × 8 | 8 × 8 | 8 × 8 | 8 × 8 | 8 × 10 | 5 |
| 15 to 20 | Up to 6 | 6 × 8 | 6 × 8 | 6 × 8 | 8 × 8 | 8 × 8 | 5 |
| | Up to 8 | 8 × 8 | 8 × 8 | 8 × 8 | 8 × 8 | 8 × 10 | 5 |
| | Up to 10 | 8 × 10 | 8 × 10 | 8 × 10 | 8 × 10 | 10 × 10 | 5 |

# TIMBER TRENCH SHORING MINIMUM TIMBER REQUIREMENTS — TYPE B SOIL (cont.)

$P(a) = 45 \times H + 72$ psf (2 ft. surcharge)

| Depth of Trench (feet) | Wales | | Uprights | | |
|---|---|---|---|---|---|
| | Size (inches) | Vertical Spacing (feet) | Maximum Allowable Horizontal Spacing (feet) | | |
| | | | Close | 2 | 3 |
| 5 to 10 | 6 × 8 | 5 | – | – | 2 × 6 |
| | 8 × 10 | 5 | – | – | 2 × 6 |
| | 10 × 10 | 5 | – | – | 2 × 6 |
| 10 to 15 | 8 × 8 | 5 | – | 2 × 6 | – |
| | 10 × 10 | 5 | – | 2 × 6 | – |
| | 10 × 12 | 5 | – | 2 × 6 | – |
| 15 to 20 | 8 × 10 | 5 | 3 × 6 | – | – |
| | 10 × 12 | 5 | 3 × 6 | – | – |
| | 12 × 12 | 5 | 3 × 6 | – | – |

Mixed oak or equivalent with a bending strength not less than 850 psi. Manufactured members of equivalent strength may be substituted for wood.

# TIMBER TRENCH SHORING MINIMUM TIMBER REQUIREMENTS — TYPE B SOIL *(cont.)*

$$P(a) = 45 \times H + 72 \text{ psf (2 ft. surcharge)}$$

| Depth of Trench (feet) | Horiz. Spacing (feet) | Size (S4S) and Spacing of Members | | | | | Vertical Spacing (feet) |
|---|---|---|---|---|---|---|---|
| | | Cross Braces | | | | | |
| | | Width of Trench (feet) | | | | | |
| | | Up to 4 | Up to 6 | Up to 9 | Up to 12 | Up to 15 | |
| 5 to 10 | Up to 6 | 4 × 6 | 4 × 6 | 4 × 6 | 6 × 6 | 6 × 6 | 5 |
| | Up to 8 | 4 × 6 | 4 × 6 | 6 × 6 | 6 × 6 | 6 × 6 | 5 |
| | Up to 10 | 4 × 6 | 4 × 6 | 6 × 6 | 6 × 6 | 6 × 8 | 5 |
| 10 to 15 | Up to 6 | 6 × 6 | 6 × 6 | 6 × 6 | 6 × 8 | 6 × 8 | 5 |
| | Up to 8 | 6 × 8 | 6 × 8 | 6 × 8 | 8 × 8 | 8 × 8 | 5 |
| | Up to 10 | 6 × 8 | 6 × 8 | 8 × 8 | 8 × 8 | 8 × 8 | 5 |
| 15 to 20 | Up to 6 | 6 × 8 | 6 × 8 | 6 × 8 | 6 × 8 | 8 × 8 | 5 |
| | Up to 8 | 6 × 8 | 6 × 8 | 6 × 8 | 8 × 8 | 8 × 8 | 5 |
| | Up to 10 | 8 × 8 | 8 × 8 | 8 × 8 | 8 × 8 | 8 × 8 | 5 |

## TIMBER TRENCH SHORING MINIMUM TIMBER REQUIREMENTS — TYPE B SOIL (cont.)

P(a) = 45 × H + 72 psf (2 ft. surcharge)

| Depth of Trench (feet) | Wales | | Uprights | | |
|---|---|---|---|---|---|
| | Size (inches) | Vertical Spacing (feet) | Maximum Allowable Horizontal Spacing (feet) | | |
| | | | Close | 2 | 3 |
| 5 to 10 | 6 × 8 | 5 | – | – | 3 × 12<br>4 × 8 |
| | 8 × 8 | 5 | – | 3 × 8 | |
| | 8 × 10 | 5 | – | – | 4 × 8 |
| 10 to 15 | 8 × 8 | 5 | 3 × 6 | 4 × 10 | – |
| | 10 × 10 | 5 | 3 × 6 | 4 × 10 | – |
| | 10 × 12 | 5 | 3 × 6 | 4 × 10 | – |
| 15 to 20 | 8 × 10 | 5 | 4 × 6 | – | – |
| | 10 × 12 | 5 | 4 × 6 | – | – |
| | 12 × 12 | 5 | 4 × 6 | – | – |

Douglas fir or equivalent with a bending strength not less than 1500 psi. Manufactured members of equivalent strength may be substituted for wood.

# TIMBER TRENCH SHORING MINIMUM TIMBER REQUIREMENTS — TYPE C SOIL

$P(a) = 80 \times H + 72$ psf (2 ft. surcharge)

| Depth of Trench (feet) | Size (actual) and Spacing of Members | | | | | | |
|---|---|---|---|---|---|---|---|
| | Cross Braces | | | | | | |
| | Horiz. Spacing (feet) | Width of Trench (feet) | | | | | Vertical Spacing (feet) |
| | | Up to 4 | Up to 6 | Up to 9 | Up to 12 | Up to 15 | |
| 5 to 10 | Up to 6 | 6 × 8 | 6 × 8 | 6 × 8 | 8 × 8 | 8 × 8 | 5 |
| | Up to 8 | 8 × 8 | 8 × 8 | 8 × 8 | 8 × 8 | 8 × 10 | 5 |
| | Up to 10 | 8 × 10 | 8 × 10 | 8 × 10 | 8 × 10 | 10 × 10 | 5 |
| 10 to 15 | Up to 6 | 8 × 8 | 8 × 8 | 8 × 8 | 8 × 8 | 8 × 10 | 5 |
| | Up to 8 | 8 × 10 | 8 × 10 | 8 × 10 | 8 × 10 | 10 × 10 | 5 |
| 15 to 20 | Up to 6 | 8 × 10 | 8 × 10 | 8 × 10 | 8 × 10 | 10 × 10 | 5 |

# TIMBER TRENCH SHORING MINIMUM TIMBER REQUIREMENTS — TYPE C SOIL *(cont.)*

$P(a) = 80 \times H + 72$ psf (2 ft. surcharge)

| Depth of Trench (feet) | Size (actual) and Spacing of Members | | Uprights |
|---|---|---|---|
| | Wales | | Maximum Allowable Horizontal Spacing (feet) |
| | Size (inches) | Vertical Spacing (feet) | Close |
| 5 to 10 | 8 × 10 | 5 | 2 × 6 |
| | 10 × 12 | 5 | 2 × 6 |
| | 12 × 12 | 5 | 2 × 6 |
| 10 to 15 | 10 × 12 | 5 | 2 × 6 |
| | 12 × 12 | 5 | 2 × 6 |
| 15 to 20 | 12 × 12 | 5 | 3 × 6 |

Mixed oak or equivalent with a bending strength not less than 850 psi. Manufactured members of equivalent strength may be substituted for wood.

# TIMBER TRENCH SHORING MINIMUM TIMBER REQUIREMENTS — TYPE C SOIL *(cont.)*

P(a) = 80 × H + 72 psf (2 ft. surcharge)

| Depth of Trench (feet) | Horiz. Spacing (feet) | Size (S4S) and Spacing of Members | | | | | Vertical Spacing (feet) |
|---|---|---|---|---|---|---|---|
| | | Cross Braces | | | | | |
| | | Width of Trench (feet) | | | | | |
| | | Up to 4 | Up to 6 | Up to 9 | Up to 12 | Up to 15 | |
| 5 to 10 | Up to 6 | 6 × 6 | 6 × 6 | 6 × 6 | 6 × 6 | 8 × 8 | 5 |
| | Up to 8 | 6 × 6 | 6 × 6 | 6 × 6 | 8 × 8 | 8 × 8 | 5 |
| | Up to 10 | 6 × 6 | 6 × 6 | 8 × 8 | 8 × 8 | 8 × 8 | 5 |
| 10 to 15 | Up to 6 | 6 × 8 | 6 × 8 | 6 × 8 | 8 × 8 | 8 × 8 | 5 |
| | Up to 8 | 8 × 8 | 8 × 8 | 8 × 8 | 8 × 8 | 8 × 8 | 5 |
| 15 to 20 | Up to 6 | 8 × 8 | 8 × 8 | 8 × 8 | 8 × 10 | 8 × 10 | 5 |

## TIMBER TRENCH SHORING MINIMUM TIMBER REQUIREMENTS — TYPE C SOIL (cont.)

$P(a) = 80 \times H + 72$ psf (2 ft. surcharge)

| Depth of Trench (feet) | Size (S4S) and Spacing of Members | | Uprights |
| | Wales | | |
| | Size (inches) | Vertical Spacing (feet) | Maximum Allowable Horizontal Spacing (feet) |
| | | | Close |
|---|---|---|---|
| 5 to 10 | $8 \times 8$ | 5 | $3 \times 6$ |
| | $10 \times 10$ | 5 | $3 \times 6$ |
| | $10 \times 12$ | 5 | $3 \times 6$ |
| 10 to 15 | $10 \times 10$ | 5 | $4 \times 6$ |
| | $12 \times 12$ | 5 | $4 \times 6$ |
| 15 to 20 | $10 \times 12$ | 5 | $4 \times 6$ |

## LOAD-CARRYING CAPACITIES OF SOILS

| Type of Soil | Tons per sq. ft. |
|---|---|
| Soft clay | 1 |
| Firm clay or fine sand | 2 |
| Compact fine or loose coarse sand | 3 |
| Loose gravel or compact coarse sand | 4 |
| Compact sand-gravel mixture | 6 |

# A WELLPOINT SYSTEM DEPRESSING THE WATER TABLE

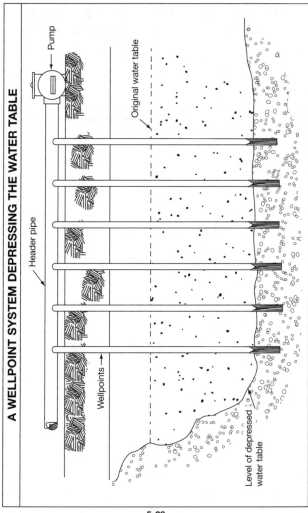

Pump

Header pipe

Original water table

Wellpoints

Level of depressed water table

# CHAPTER 6
## *Electrical Safety*

### BASIC ELECTRICAL SAFETY GUIDELINES

- Always comply with the National Electric Code.
- Use UL-approved appliances, components and equipment.
- Keep electrical grounding circuits in good condition. Ground any conductive component or element that does not have to be energized. The grounding connection must be a low-resistance conductor heavy enough to carry the largest fault current that may occur.
- Turn OFF, lock out and tag disconnect switches when working on any electrical circuit or equipment. Test all circuits after they are turned OFF.
- Use double-insulated power tools or power tools that include a third conductor grounding terminal, which provides a path for fault current.
- Always use protective and safety equipment.
- Check conductors, cords, components and equipment for signs of wear or damage.
- Never throw water on an electrical fire. Turn OFF the power and use a Class C–rated fire extinguisher.
- Never work alone when working in a dangerous area or with dangerous equipment.
- Learn CPR and first aid.
- Do not work in poorly lighted areas.

## BASIC ELECTRICAL SAFETY GUIDELINES *(cont.)*

- Always use nonconductive ladders. Never use a metal ladder.
- Ensure there are no atmospheric hazards, such as flammable dust or vapor, in the area.
- Use one hand when working on a live circuit to reduce the chance of an electrical shock passing through the heart and lungs.
- Never bypass or disable fuses or circuit breakers.

## EFFECT OF ELECTRIC CURRENT ON HUMAN BODY

| Current (ma) | Effect | Result |
|---|---|---|
| 0 to 6 | Slight sensation possible | None |
| 6 to 15 | Painful shock muscular contraction | Possible No "Let Go" |
| 15 to 20 | Painful shock Frozen until circuit is de-energized | No "Let Go" |
| 20 to 50 | Severe muscular contractions Asphyxia | Often Fatal |
| 50 to 200 | Ventricular fibrillation | Probably Fatal |
| 2001 | Heart movement stops | Fatal |

1 ma = 0.001 amp
Duration of current flow is not noted in this comparison

## LOCKOUT/TAGOUT GUIDELINES

Electrical power must be turned off when any type of electrical equipment is inspected, serviced, repaired or replaced. The equipment must also be locked out and tagged out to ensure the safety of personnel working with the equipment.

Lockout is the process of turning off the source of electrical power and installing a lock that prevents the power from being turned on. Tagout is the process of placing a danger tag on the source of electrical power that indicates the equipment may not be operated until the danger tag is removed.

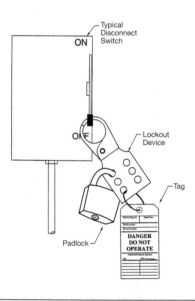

## LOCKOUT/TAGOUT TRAINING

- The purpose for lockout and tagout procedures must be understood by all appropriate personnel.
- Training in the hazards involved while performing a lockout and tagout procedure must be understood by all appropriate personnel.
- Training in the hazards of working on energized equipment and the procedures concerning this work must be understood by all appropriate personnel.

## LOCKOUT/TAGOUT RULES

- Only authorized personnel are permitted to lockout equipment.
- Locks and tags shall be uniform and easily identified as a lockout or tagout and only used to lockout equipment for personal protection.
- Lockout tags will accompany all locks or be used as the sole means of lockout when equipment is configured without a locking mechanism. When lockout tags are used only, they shall be uniform and easily identified as a tagout to be used for personal protection.
- No one other than the person who installed the lockout tag-out device shall attempt to operate any switch, valve or isolating device bearing a lock or lockout tag.
- Exception: The employer may remove a lock or tag if the employee who placed the locking device is unable to remove it due to injury, sickness or other unavailablility.

## LOCK REMOVAL RULES

- Verify that the authorized employee is incapacitated or not at the facility.
- Make all reasonable efforts to contact the authorized employee to inform him/her that his/her lock and tag were removed.
- Ensure that the authorized employee has this knowledge before he/she resumes work at the facility.
- All possible means must be used to contact the employee of the removal of his/her lockout device before the employee returns to the job site. (This situation may provide for an undesirable condition if the employee returns to the job site expecting the equipment to be locked out.)

## WORK AREA CONTROL

- A barrier tape, cones, fence or other methods may be used to surround any work area where a hazard exists that could be accessed by personnel unfamiliar with that work, such as on-lookers.
- The tape shall be prominently placed and completely surround the area where the hazard exists.
- The taped area shall be large enough to permit all persons working within the work area adequate clearance from any hazard.
- Remove all barriers when the hazard is eliminated.
- Metal barrier tape shall not be used.
- There may be a need for attendants to provide additional safety for personnel protection. This may include the use of personnel with flags, detour signs, cones, etc.

| MINIMUM DEPTH WORKING CLEARANCES | | | |
|---|---|---|---|
| Nominal Voltage to Ground | Working Conditions (feet) | | |
| | 1 | 2 | 3 |
| 0–150 | 3 | 3 | 3 |
| 151–600 | 3 | 3½ | 4 |
| 601–2,500 | 3 | 4 | 5 |
| 2,501–9,000 | 4 | 5 | 6 |
| 9,001–25,000 | 5 | 6 | 9 |
| 25,001–75,000 | 6 | 8 | 10 |
| Above 75,000 | 8 | 10 | 12 |

| ELEVATION OF UNGUARDED LIVE PARTS ABOVE WORKING SPACE | |
|---|---|
| Nominal Voltage Between Phases | Elevation (feet) |
| 601–7,500 | 9 |
| 7,501–35,000 | 9½ |
| Over 35,000 | 9½' + 0.37" for every 1,000 volts over 35,000 volts |

## MINIMUM CLEARANCE — LIVE PARTS

| *Minimum Clearance Of Live Parts (inches) | | | | Nominal Voltage Rating (kV) | Impulse Withstand B.I.L. (kV) | |
|---|---|---|---|---|---|---|
| Phase-To-Phase | | Phase-To-Ground | | | | |
| Indoors | Outdoors | Indoors | Outdoors | | Indoors | Outdoors |
| 4.5 | 7 | 3.0 | 6 | 2.4–4.16 | 60 | 95 |
| 5.5 | 7 | 4.0 | 6 | 7.2 | 75 | 95 |
| 7.5 | 12 | 5.0 | 7 | 13.8 | 95 | 110 |
| 9.0 | 12 | 6.5 | 7 | 14.4 | 110 | 110 |
| 10.5 | 15 | 7.5 | 10 | 23 | 125 | 150 |
| 12.5 | 15 | 9.5 | 10 | 34.5 | 150 | 150 |
| 18.0 | 18 | 13.0 | 13 | — | 200 | 200 |
| — | 18 | — | 13 | 46 | — | 200 |
| — | 21 | — | 17 | — | — | 250 |
| — | 21 | — | 17 | 69 | — | 250 |
| — | 31 | — | 25 | — | — | 350 |
| — | 53 | — | 42 | 115 | — | 550 |
| — | 53 | — | 42 | 138 | — | 550 |
| — | 63 | — | 50 | — | — | 650 |
| — | 63 | — | 50 | 161 | — | 650 |
| — | 72 | — | 58 | — | — | 750 |
| — | 72 | — | 58 | 230 | — | 750 |
| — | 89 | — | 71 | — | — | 900 |
| — | 105 | — | 83 | — | — | 1050 |

This represents the minimum clearance for rigid parts and bare conductors under favorable conditions. They shall be increased for conductor movement, unfavorable conditions or where space is limited. Impulse withstand voltage for a particular system voltage is determined by the type of surge protection equipment that is utilized.

# UNDERGROUND
## INSTALLATION REQUIREMENTS

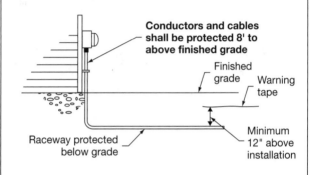

**Conductors and cables shall be protected 8' to above finished grade**

Finished grade

Warning tape

Raceway protected below grade

Minimum 12" above installation

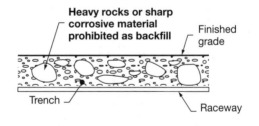

**Heavy rocks or sharp corrosive material prohibited as backfill**

Finished grade

Trench

Raceway

# UNDERGROUND
## INSTALLATION REQUIREMENTS *(cont.)*

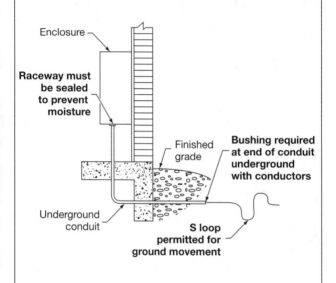

Enclosure

**Raceway must
be sealed
to prevent
moisture**

Finished
grade

**Bushing required
at end of conduit
underground
with conductors**

Underground
conduit

**S loop
permitted for
ground movement**

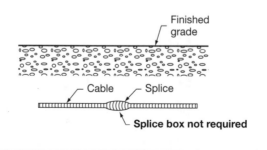

Finished
grade

Cable — Splice

**Splice box not required**

## MINIMUM COVER REQUIREMENTS FOR UNDERGROUND INSTALLATIONS

| Location of circuit or wiring method | Type of circuit or wiring method (0 to 600 Volts, nominal) | | | | |
|---|---|---|---|---|---|
| | Direct burial cables or conductors | Rigid metal conduit or intermediate metal conduit | Non-metallic raceways listed for direct burial without concrete encasement or other approved raceways | Residential branch circuits rated 120 V or less with GFCI protection and maximum over-current protection of 20 amperes | Circuits for control of irrigation and landscape lighting limited to not more than 30 V and installed with type UF or in other identified cable or raceway |
| All locations not specified in these charts | 24" | 6" | 18" | 12" | 6" |
| In trench below 2" thick concrete or equivalent | 18" | 6" | 12" | 6" | 6" |
| Under buildings | — (In raceway only) | — | — | — (In raceway only) | — (In raceway only) |
| Under minimum of 4" thick concrete exterior slab with no traffic and slab extending no less than 6" beyond the installation | 18" | 4" | 4" | 6" (Direct burial) 4" (In raceway) | 6" (Direct burial) 4" (In raceway) |
| Under alleys, highways, roads, driveways, streets, and parking lots | 24" | 24" | 24" | 24" | 24" |

6-10

# MINIMUM COVER REQUIREMENTS FOR UNDERGROUND INSTALLATIONS (cont.)

| Location of circuit or wiring method | Type of circuit or wiring method (0 to 600 Volts, nominal) | | | | |
|---|---|---|---|---|---|
| | Direct burial cables or conductors | Rigid metal conduit or intermediate metal conduit | Non-metallic raceways listed for direct burial without concrete encasement or other approved raceways | Residential branch circuits rated 120 V or less with GFCI protection and maximum over-current protection of 20 amperes | Circuits for control of irrigation and landscape lighting limited to not more than 30 V and installed with type UF or in other identified cable or raceway |
| Under family dwelling driveways. Outdoor parking areas used for dwelling-related purposes only | 18" | 18" | 18" | 12" | 18" |
| In or under airport runways, including areas where the public is prohibited | 18" | 18" | 18" | 18" | 18" |

Notes: Cover is defined as the shortest distance between a point on the top service of any direct-buried conductor, cable, conduit or other raceway and the top surface of finished grade, concrete, etc.

Raceways approved for burial only where concrete encased shall require a 2" concrete envelope.

Lesser depths are permitted where cables/conductors rise for terminations, splices or equipment.

Where solid rock prevents compliance with the cover depths specified in this table, the wiring must be installed in raceways permitted for direct burial. The raceways shall be covered by a minimum of 2" of concrete extending down to the solid rock.

## BASIC GROUNDED CONDUCTOR RULES

**Circuit breakers or switches shall not disconnect the grounded conductor of a circuit.**

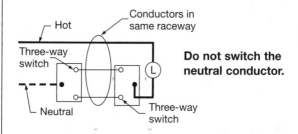

Conductors in same raceway

Hot

Three-way switch

**Do not switch the neutral conductor.**

Neutral

Three-way switch

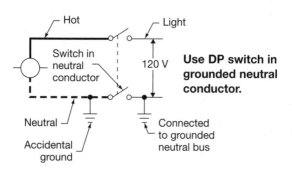

Hot

Light

Switch in neutral conductor

120 V

**Use DP switch in grounded neutral conductor.**

Neutral

Accidental ground

Connected to grounded neutral bus

**Exception**: A circuit breaker or switch may disconnect grounded circuit conductor if all circuit conductors are disconnected at the same time.

## BASIC GROUNDED CONDUCTOR RULES *(cont.)*

**Circuit breakers or switches shall not disconnect the grounded conductor of a circuit.**

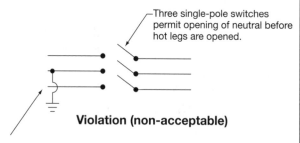

Three single-pole switches permit opening of neutral before hot legs are opened.

**Violation (non-acceptable)**

Three-wire, 1φ supply

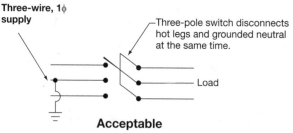

Three-pole switch disconnects hot legs and grounded neutral at the same time.

Load

**Acceptable**

**Exception:** A circuit breaker or switch may disconnect the grounded circuit conductor if it cannot be disconnected until all other ungrounded conductors have been disconnected.

# BASIC GROUNDING CONNECTIONS

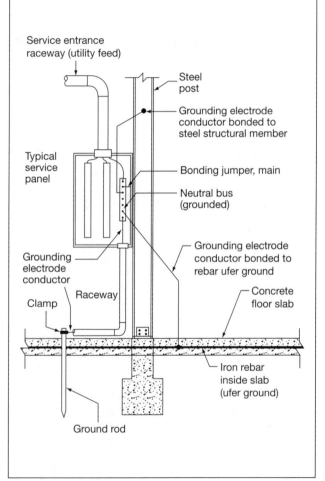

Service entrance raceway (utility feed)

Steel post

Grounding electrode conductor bonded to steel structural member

Typical service panel

Bonding jumper, main

Neutral bus (grounded)

Grounding electrode conductor bonded to rebar ufer ground

Grounding electrode conductor

Concrete floor slab

Clamp

Raceway

Iron rebar inside slab (ufer ground)

Ground rod

6-14

## MINIMUM SIZES OF GROUNDING ELECTRODE CONDUCTORS FOR AC SYSTEMS

| Size of Largest Service-Entrance Conductor or Equivalent for Parallel Conductors | | Size of Grounding Electrode Conductor | |
|---|---|---|---|
| Copper | Aluminum or Copper-Clad Aluminum | Copper | Aluminum or Copper-Clad Aluminum |
| 2 or smaller | 0 or smaller | 8 | 6 |
| 1 or 2 | 2/0 or 3/0 | 6 | 4 |
| 2/0 or 3/0 | 4/0 or 250 kcmil | 4 | 2 |
| Over 3/0 through 350 kcmil | Over 250 kcmil through 500 kcmil | 2 | 0 |
| Over 350 kcmil through 600 kcmil | Over 500 kcmil through 900 kcmil | 0 | 3/0 |
| Over 600 kcmil through 1100 kcmil | Over 900 kcmil through 1750 kcmil | 2/0 | 4/0 |
| Over 1100 kcmil | Over 1750 kcmil | 3/0 | 250 kcmil |

## RESISTIVITIES OF VARIOUS SOILS

| Types of Soil | Resistivity (ohms per cubic meter) | | |
|---|---|---|---|
| | Average | Min. | Max. |
| **WASTE** Fills—ashes, cinders, brine | 2,370 | 590 | 7,000 |
| Clay, shale, gumbo, loam | 4,060 | 340 | 16,300 |
| Same—with varying proportions of sand and gravel | 15,800 | 1,020 | 135,000 |
| Gravel, sand, stones, with little clay or loam | 94,000 | 59,000 | 458,000 |

# MINIMUM SIZE CONDUCTORS — GROUNDING RACEWAY AND EQUIPMENT

| Conductor Size | | Amperage Rating or Setting of Automatic Overcurrent Device Not to Exceed |
|---|---|---|
| Copper | Aluminum or Copper-Clad Aluminum | |
| 14 AWG | 12 AWG | 15 amps |
| 12 | 10 | 20 amps |
| 10 | 8 | 30 amps |
| 10 | 8 | 40 amps |
| 10 | 8 | 60 amps |
| 8 | 6 | 100 amps |
| 6 | 4 | 200 amps |
| 4 | 2 | 300 amps |
| 3 | 1 | 400 amps |
| 2 | 1/0 | 500 amps |
| 1 | 2/0 | 600 amps |
| 1/0 | 3/0 | 800 amps |
| 2/0 | 4/0 | 1000 amps |
| 3/0 | 250 kcmil | 1200 amps |
| 4/0 | 350 | 1600 amps |
| 250 kcmil | 400 | 2000 amps |
| 350 | 600 | 2500 amps |
| 400 | 600 | 3000 amps |
| 500 | 800 | 4000 amps |
| 700 | 1200 | 5000 amps |
| 800 | 1200 | 6000 amps |

The equipment grounding conductor shall be sized larger than this table per National Electric Code installation restrictions.

TYPES OF GROUNDING METHODS PER THE NEC

Below ground metal water supply pipe

10' Min

Ground lug

Metal frame

Concrete-encased electrode #4 or larger conductor minimum 20' in length

Metal building frame

Ground ring — #2 bare copper conductor minimum 20' of length

30" or more

6-17

# GROUNDING DIFFERENT TYPES OF CIRCUITS

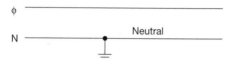

**Single-Phase, Two-Wire**

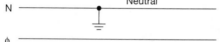

**Single-Phase, Three-Wire**

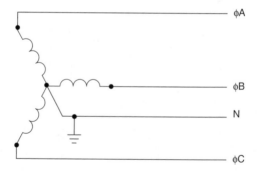

**Three-Phase Wye**

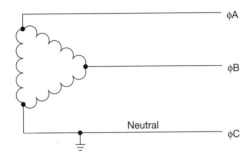

φA

φB

Neutral

φC

**Three-Phase Delta, Three-Wire**

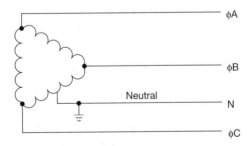

φA

φB

Neutral

N

φC

**Three-Phase Delta, Four-Wire**

# GFCI WIRING DIAGRAMS

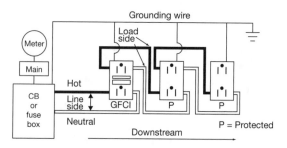

## Feed-Thru Installation

To protect the entire branch circuit, the GFCI must be the first receptacle from the circuit breaker or fuse box. Receptacles on the circuit downstream from the GFCI will also be protected.

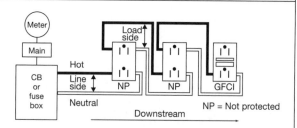

## Non-Feed-Thru Installation on a Two-Wire Circuit

Terminal protection can be achieved on a multi-outlet circuit by connecting the hot and neutral line conductors to the corresponding line side terminals of the GFCI. Only the GFCI receptacle will be protected.

## CLASSIFICATIONS OF HAZARDOUS LOCATIONS

| Classes | Likelihood that a flammable or combustible concentration is present |
|---------|---------------------------------------------------------------------|
| I | Sufficient quantities of flammable gases and vapors present in air to cause an explosion or ignite hazardous materials. |
| II | Sufficient quantities of combustible dust are present in air to cause an explosion or ignite hazardous materials. |
| III | Easily ignitable fibers or flyings are present in air, but not in a sufficient quantity to cause an explosion or ignite hazardous materials. |

| Divisions | Location containing hazardous substances |
|-----------|------------------------------------------|
| 1 | Hazardous location in which hazardous substance is **normally present** in air in sufficient quantities to cause an explosion or ignite hazardous materials. |
| 2 | Hazardous location in which hazardous substance is **not normally present** in air in sufficient quantities to cause an explosion or ignite hazardous materials. |

| Groups | Atmosphere containing flammable gases or vapors or combustible dust | |
|--------|---------|----------|
| **Class I** | **Class II** | **Class III** |
| A | E | |
| B | F | none |
| C | G | |
| D | | |

# STANDARD AND OPTIONAL
# HAZARDOUS LOCATION SYSTEMS

## Standard System

**Class I**         **Flammable Gases or Vapors**
Division 1          Normally Present in Air
Division 2          Not Normally Present in Air

**Class II**        **Combustible Dust**
Division 1          Normally Present in Air
Division 2          Not Normally Present in Air

**Class III**       **Ignitable Fibers or Flyings**
Division 1          Normally Present in Air
Division 2          Not Normally Present in Air

## Optional System*

**Zone 0**          Flammable Gases or Vapors are Present
                    Continuously or for Long Periods of Time
**Zone 1**          Flammable Gases or Vapors are Likely to Exist Under
                    Normal Operating Conditions or Exist Frequently
**Zone 2**          Flammable Gases or Vapors are not Normally
                    Present or are Present for Short Periods of Time

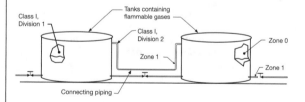

*The optional classification system as well as area classifications,
wiring methods and equipment selection can only be utilized under
supervision of a qualified registered professional engineer.

# CHAPTER 7
## *Lamp Disposal*

Upgrading a lighting system almost always involves the removal and disposal of lamps and ballasts. Some of this waste may be considered hazardous, and you must manage it accordingly. Your company may face severe penalties if you do not.

### PCB-CONTAINING BALLASTS

The primary concern regarding the disposal of used fluorescent ballasts is the health risk associated with polychlorinated biphenyls (PCBs), the same chemical that necessitated the removal of many oil-filled transformers in recent years. Fluorescent and high-intensity discharge (HID) ballasts use a small capacitor that may contain high concentrations of PCBs. The Toxic Substances Control Act (TSCA), enacted in 1976, subsequently banned the production of PCBs in the United States.

The proper method for disposing used ballasts depends on several factors, such as the type and condition of the ballasts and the regulations or recommendations in effect in the state(s) where you remove or discard them. TSCA specifies the disposal method for ballasts that are *leaking* PCBs. In addition, generators of PCB-containing ballast wastes may be subject to notification and liability provisions under the Comprehensive Environmental Response, Compensation and Liability Act of 1980 (CERCLA) also known as "Superfund." Because disposal requirements vary from state to state, check with regional, state or local authorities for all applicable regulations in your area.

## IDENTIFYING PCB BALLASTS

Use the following guidelines to identify ballasts that contain PCBs:

- All ballasts manufactured through 1979 contain PCBs.
- Ballasts manufactured after 1979 that do not contain PCBs are labeled "No PCBs."
- If a ballast is not labeled "No PCBs," assume it contains PCBs.
- It is extremely important to find out whether a ballast containing PCBs is leaking before you remove it from the fixture, so that you can handle it properly. If you do not, it can get expensive.

## NON-LEAKING PCB BALLAST DISPOSAL

Under federal laws, intact fluorescent and HID ballasts that are *not* leaking PCBs may be disposed in a municipal solid waste landfill. The Environmental Protection Agency (EPA) recommends, but cannot demand, packing and sealing the intact ballasts in 55 gallon drums. Green Lights recommends high-temperature incineration, recycling or disposal in a chemical or hazardous waste landfill.

In addition, the federal laws regulate the disposal of non-leaking PCB-containing ballasts, requiring

## NON-LEAKING PCB BALLAST DISPOSAL *(cont.)*

building owners and waste generators to notify the National Response Center at (800) 424-8802. They must notify when disposing a pound or more of PCBs (roughly equivalent to 12–16 fluorescent ballasts) in a 24-hour period.

A note of warning: The federal government has a history of applying environmental laws after the fact (ex post facto), so as a generator of PCB-containing ballast wastes, you could be liable in a subsequent cleanup at a municipal, hazardous or chemical land disposal site, incinerator or recycling facility.

## LEAKING PCB BALLASTS

A puncture or other damage to ballasts in a lighting system exposes an oily, tar-like substance. If this substance contains PCBs, the ballast and all materials it contacts are considered PCB waste and are subject to federal requirements. Leaking PCB-containing ballasts must be incinerated at an EPA-approved high-temperature incinerator. Take precau-tions to prevent exposure of the leaking ballast, because all materials that contact the ballast or the leaking substance are also PCB waste. Use trained personnel or contractors to handle and dispose leaking PCB-containing ballasts. For proper packing, storage, transportation and disposal information, call the TSCA assistance information hotline at (202) 554-1404.

## STATE REQUIREMENTS FOR PCB BALLASTS

Many states have developed regulations governing the disposal of PCB-containing ballasts that are more stringent than federal regulations. In addition, some EPA Regional offices have published policies specifying ballast disposal methods adopted by individual states.

State standards can take several forms (written regulations, regional policies, written and verbal recommendations, transportation documentation). Some states do not regulate PCB-containing ballasts as toxic waste, but prohibit their disposal in municipal solid waste landfills.

All generators of PCB-containing ballasts should thoroughly investigate their state's regulations and follow local requirements. There are three common methods for disposing of non-leaking PCB-containing ballasts: high-temperature incineration, recycling and chemical or hazardous waste landfill.

## HIGH-TEMPERATURE INCINERATION OF PCB BALLASTS

High-temperature incineration is the method preferred by many companies because it destroys PCBs, removing them from the waste stream permanently and removing the potential for future liability. Incinerating a PCB-containing ballast costs more than sending it to a hazardous waste landfill, but this additional cost is one many organizations are willing to absorb.

## RECYCLING BALLASTS

Recyclers remove the PCB-containing materials (the capacitor and possibly the asphalt potting material surrounding the capacitor) for incineration or land disposal. Metals such as copper and steel can be reclaimed from the ballasts for use in manufacturing other products.

You may recycle used non-leaking ballasts despite PCBs.

## DISPOSING PCB BALLASTS AT A CHEMICAL OR HAZARDOUS WASTE LANDFILL

PCB-containing ballasts may also be disposed in a chemical or hazardous waste landfill. Landfill disposal is less expensive than high-temperature incineration or recycling, but does not eliminate PCBs from the waste stream permanently. Although chemical or hazardous waste landfill disposal is an acceptable, regulated disposal method, you may be legitimately concerned about potential future liability using this method.

## PACKING PCB BALLASTS FOR DISPOSAL

Despite the disposal method selected, ballasts are packed according to PCB regulations in 55-gallon drums for transportation. One drum holds 150 to 300 ballasts depending on how tightly the ballasts are packed. Fill the void spaces with an absorbent packing material for safety reasons, and label the drums according to Department of Transportation regulations.

## PACKING PCB BALLASTS FOR DISPOSAL *(cont.)*

Note that tightly packed drums may weigh more than 1,000 pounds, which may present a safety risk, particularly when moving the drum for loading or unloading.

## DISPOSAL COSTS FOR PCB BALLASTS

High-temperature incineration and chemical or hazardous waste landfill costs can vary considerably. Disposal prices vary according to:

- Quantity of waste generated.
- Location of removal site.
- Proximity to an EPA-approved high-temperature incinerator or chemical or hazardous waste landfill.
- State and local taxes.

When shopping for ballast disposal services, request cost estimates in terms of both pounds and number of ballasts. Typical F40 ballasts weigh about 3.5 lbs., and F96 ballasts weigh about 8 lbs. Negotiate with hazardous waste brokers, transporters, waste management companies and disposal sites to obtain the lowest fees.

Note that these are average price figures, and you should verify prices in your area before estimating your costs.

## HIGH-TEMPERATURE INCINERATION COSTS FOR PCB BALLASTS

Incineration costs are calculated by weight. Costs range from $0.55/lb. to $2.10/lb. Average cost is $1.50/lb., which equals approximately $5.25 per ballast. Note: Estimated costs do not include packaging, transportation or profile fees.

## RECYCLING COSTS FOR PCB BALLASTS

When recyclers remove the PCB-containing capacitor, the volume and weight of the ballast are reduced. This change results in lower packing, transportation and incineration or disposal costs. Recycling costs are calculated by weight. Costs range from $0.75/lb. to $1.75/lb. Average cost is $1.00/lb., which equals approximately $3.50 per ballast. Note: Recycling cost can range from $1.25 per ballast (if the PCB wastes are sent to a chemical or hazardous waste landfill) to approximately $3.50 per ballast (if the PCB wastes are high-temperature incinerated). Estimated costs do not include packaging, transportation or profile fees.

## DISPOSAL COSTS FOR PCB BALLASTS AT A CHEMICAL OR HAZARDOUS WASTE LANDFILL

Chemical or hazardous waste landfill costs are calculated per 55-gallon drum. Costs range from $65/drum to $165/drum. Average cost is $100/drum, which equals approximately $0.50/ballast. Note: Estimated costs do not include packaging, transportation or profile fees.

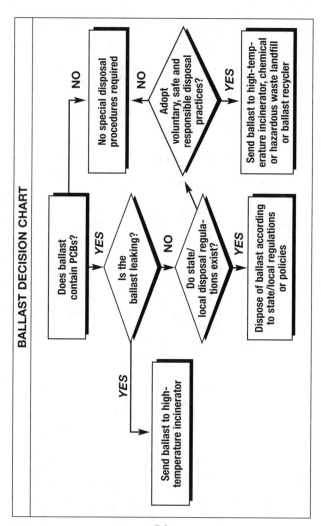

**BALLAST DECISION CHART**

Does ballast contain PCBs?

NO → No special disposal procedures required

YES → Is the ballast leaking?

YES → Send ballast to high-temperature incinerator

NO → Do state/local disposal regulations exist?

YES → Dispose of ballast according to state/local regulations or policies

Adopt voluntary, safe and responsible disposal practices?

NO → No special disposal procedures required

YES → Send ballast to high-temperature incinerator, chemical or hazardous waste landfill or ballast recycler

## TRANSPORTATION FEES

Transportation fees are calculated as cents per pound per mile. They vary according to (1) the number of drums removed from the site, and (2) the distance from your location to the location of the high-temperature incinerator, chemical or hazardous waste landfill or recycler.

Transporters may need to be registered or licensed to move hazardous wastes in certain states. Documentation of the movement of hazardous waste may be required even if a state does not regulate disposal or fees require the use of a licensed transporter.

## PROFILE FEES

Operators of the high-temperature incinerator or chemical or hazardous waste landfill may charge a profile fee to document incoming hazardous waste. Profile fees vary depending on the volume of waste materials generated. Profile fees range from $0 to $300 per delivery.

Fees may be waived if a certain volume or frequency of deliveries is assured or a working relationship has been established with a waste management broker, lighting management company or other contractor.

## RECORDKEEPING

To track transported hazardous waste, the EPA requires generators to prepare a Uniform Hazardous Waste Manifest. The hazardous waste landfill, incinerator or recycler that you use can provide this one-page form. The manifest identifies the type and quantity of waste, the generator, the transporter and its ultimate destination.

The manifest must accompany the waste wherever it travels. Each handler of the waste must sign the manifest and keep one copy. When the waste reaches its destination, the owner of that facility returns a copy of the manifest to the generator to confirm that the waste arrived. If the waste does not arrive as scheduled, generators must immediately notify EPA, or the authorized state environmental agency, so that they can investigate and act appropriately. In addition, require your contractor to provide you with documents verifying the disposal method, whether the PCBs are incinerated at high-temperatures or disposed in a chemical or hazardous waste landfill.

## DISPOSAL OF MERCURY-CONTAINING LAMPS

Fluorescent and HID lamps contain a small quantity of mercury that can be harmful to the environment and human health when improperly managed. Mercury is regulated under regulations administered by the EPA. Under current federal law, mercury-containing lamps such as fluorescent and HID lamps may or may not be considered hazardous waste. In addition, incandescent and HID lamps may contain small quantities of lead that can also be potentially harmful to human health and the environment.

The federal government requires generators of solid wastes containing mercury to determine whether the waste is hazardous by using generator knowledge or testing representative samples of that waste. According to the regulations, generators of used fluorescent and HID lamps are responsible for determining whether their lamp wastes are hazardous. If you do not test used fluorescent and HID lamps and prove them non-hazardous, assume they are hazardous waste and dispose of them accordingly.

To use generator knowledge in making a hazardous waste determination, the generator must have information about possible hazardous constituents and their quantities in the waste. Sometimes manufacturers generate solid waste as part of their manufacturing process, and can use process knowledge to determine whether the waste exhibits a characteristic of hazardous waste. However, with expired lamp wastes, the generator has little process knowledge on which to make a hazardous waste determination (because he/she is not the manufacturer). The generator could base a determination on data obtained from the manufacturer or he could refer to the EPA's study entitled *Analytical Results of Mercury in Fluorescent Lamps*.

## TESTING LAMPS

A test called the Toxicity Characteristic Leaching Procedure (TCLP) identifies whether a waste is toxic and must be managed as hazardous waste. The test attempts to replicate the conditions in a municipal landfill to detect the mercury concentration of water that would leach from the landfill.

When mercury-containing lamps are tested using the TCLP, the test results can vary considerably, depending on the lamp manufacturer, age of the lamp and the laboratory procedures used. These lamps often fail the TCLP. If you do not use the TCLP to verify that your lamps are non-hazardous, you should (1) assume that they are hazardous waste and (2) manage them as hazardous waste. Contact your state hazardous waste agency for information about laboratories in your state that conduct the TCLP. The cost to test one lamp is usually about $140.

There is exemption from some disposal requirements if you or your customers qualify as a *conditionally exempt small quantity generator*, which is defined as a generator that disposes 100 kg or less of hazardous waste per month. Generators must add the weight of all the hazardous waste (lamps plus other hazardous wastes) that their business generates during a month. For lamp disposal, this quantity of waste includes the mercury in the lamp along with the glass, phosphors and other materials (the weight of the entire lamp). Conditionally exempt small-quantity generators are excused from identification, storage, treatment and disposal regulations. To qualify as a conditionally exempt small-quantity generator (if the only hazardous waste is mercury-containing lamps), a generator must dispose of fewer than 300–350 four-foot T12 fluorescent lamps or 400–450 4' T8 fluorescent lamps per month, depending on the approximate weight of each lamp.

## LAMP RECYCLING COSTS

Recycling costs for fluorescent lamps are typically calculated by linear foot. HID lamp recycling costs are typically quoted on a per-lamp basis. Fluorescent recycling costs range from $0.06/ft. to $0.15/ft., with the average cost being $0.10/ft. Approximately $0.40 per F40 lamp. HID recycling costs range from $1.25/lamp to $4.50/lamp, average cost is $2.50/lamp. Note: Estimated costs do not include packaging, transportation or profile fees.

## DISPOSAL COSTS FOR LAMPS AT A CHEMICAL OR HAZARDOUS WASTE LANDFILL

Disposal costs for fluorescent lamps at a hazardous waste landfill range from 25–50 cents per 4' tube, not including costs for packaging, transportation or profile fees.

## PACKING LAMPS FOR DISPOSAL

To prevent used fluorescent and HID lamps from breaking, lamps should be properly packed for storage and transportation. When lamps are removed and replaced with new lamps (during group relamping), the used lamps should be packed in the cardboard boxes that contained the replacement lamps. The boxes containing the hazardous waste must be properly labeled.

Pre-printed labels or rubber stamps that meet Department of Transportation regulations are recommended for high-volume disposal.

Small-quantity generators dispose 100 to 1,000 kg of hazardous waste per month (which roughly corresponds to 350 to 3,600 lamps) and can store hazardous waste up to 180 days.

Large-quantity generators dispose over 1,000 kg of hazardous waste per month (more than 3,600 4' lamps) and can store hazardous waste up to 90 days.

Conditionally exempt small-quantity generators dispose of 100 kg or less hazardous waste per month and are exempt from storage requirements. In addition to proper packing, care should be taken when stacking the boxes of used lamps for storage to avoid crushing the bottom boxes under the weight of the boxes on top. If you work with a contractor to maintain your lighting system, specify a safe storage arrangement in your contract. This approach ensures that your used lamps are not accidentally broken or crushed before they are sent to a disposal facility.

Some organizations crush their used lamps before disposal. This option should be pursued with care. The crushing equipment should have the approval of state and local authorities, and crushing methods should be evaluated carefully. The lamp should be crushed entirely inside the drum or storage unit so that no mercury vapor enters the atmosphere. There should also be adequate ventilation in the space where the crushing occurs. Under current EPA hazardous waste regulations, crushing lamps before sending them to a hazardous waste landfill may be considered treatment. A treatment permit may be required.

## INFORMATIONAL RESOURCES

### Postal Addresses for the
### U.S. Environmental Protection Agency

To contact individuals and offices within the U.S. Environmental Protection Agency (EPA), use the following addresses and mail codes. To find an individual's mail code, use the EPA Employee Directory.

### EPA Headquarters

**Standard Mailing Address**
Environmental Protection Agency
Ariel Rios Building
1200 Pennsylvania Avenue, NW
Mail Code 3213A
Washington, DC 20460
(202) 260-2090

**Overnight Package Delivery Mailing Address**
Environmental Protection Agency
EPA East
1201 Constitution Avenue, NW
Room number 4101 M
Washington, DC 20004

### EPA Regional Offices

**REGION 1 (CT, MA, ME, NH, RI, VT)**
Environmental Protection Agency
1 Congress Street, Suite 1100
Boston, MA 02114-2023
http://www.epa.gov/region01/
Phone: (617) 918-1111
Toll free within Region 1: (888) 372-7341
Fax: (617) 565-3660

**REGION 2 (NJ, NY, PR, VI)**
Environmental Protection Agency
290 Broadway
New York, NY 10007-1866
http://www.epa.gov/region02/
Phone: (212) 637-3000
Fax: (212) 637-3526

**REGION 3 (DC, DE, MD, PA, VA, WV)**
Environmental Protection Agency
1650 Arch Street
Philadelphia, PA 19103-2029
http://www.epa.gov/region03/
Phone: (215) 814-5000
Toll free: (800) 438-2474
Fax: (215) 814-5103
Email: r3public@epa.gov

**REGION 4 (AL, FL, GA, KY, MS, NC, SC, TN)**
Environmental Protection Agency
Atlanta Federal Center
61 Forsyth Street, SW
Atlanta, GA 30303-3104
http://www.epa.gov/region04/
Phone: (404) 562-9900
Toll free: (800) 241-1754
Fax: (404) 562-8174

**REGION 5 (IL, IN, MI, MN, OH, WI)**
Environmental Protection Agency
77 West Jackson Boulevard
Chicago, IL 60604-3507
http://www.epa.gov/region05/

## INFORMATIONAL RESOURCES *(cont.)*

Phone: (312) 353-2000
Toll free within Region 5: (800) 621-8431
Fax: (312) 353-4135

**REGION 6 (AR, LA, NM, OK, TX)**
Environmental Protection Agency
Fountain Place 12th Floor, Suite 1200
1445 Ross Avenue
Dallas, TX 75202-2733
http://www.epa.gov/region06/
Phone: (214) 665-2200
Toll free within Region 6: (800) 887-6063
Fax: (214) 665-7113

**REGION 7 (IA, KS, MO, NE)**
Environmental Protection Agency
901 North 5th Street
Kansas City, KS 66101
http://www.epa.gov/region07/
Phone: (913) 551-7003
Toll free: (800) 223-0425

**REGION 8 (CO, MT, ND, SD, UT, WY)**
Environmental Protection Agency
999 18th Street, Suite 500
Denver, CO 80202-2466
http://www.epa.gov/region08/
Phone: (303) 312-6312
Toll free: (800) 227-8917
Fax: (303) 312-6339
Email: r8eisc@epa.gov

## INFORMATIONAL RESOURCES *(cont.)*

**REGION 9 (AZ, CA, HI, NV)**
Environmental Protection Agency
75 Hawthorne Street
San Francisco, CA 94105
http://www.epa.gov/region09/
Phone: (415) 947-8000
Toll free within Region 9: (866) EPA-WEST
Fax: (415) 947-3553
Email: r9.info@epa.gov

**REGION 10 (AK, ID, OR, WA)**
Environmental Protection Agency
1200 Sixth Avenue
Seattle, WA 98101
http://www.epa.gov/region10/
Phone: (206) 553-1200
Toll free: (800) 424-4372
Fax: (206) 553-0149

### Solid and Hazardous Waste Agencies per State

**ALABAMA**
Department of Environmental Management
Land Division – Solid/Hazardous Waste
P.O. Box 301463
Montgomery, AL 36130-1463
(334) 271-7730

**ALASKA**
State of Alaska
Department of Environmental Conservation
410 Willoughby Avenue, Suite 303
Juneau, AK 99801-1795
(907) 465-5010

## INFORMATIONAL RESOURCES *(cont.)*

**ARIZONA**
Arizona Department of Environmental Quality
Hazardous Waste Permits Unit
1110 West Washington Street
Phoenix, AZ 85007
(602) 771-4153

**ARKANSAS**
Department of Environmental Quality
Hazardous Waste Division
8001 National Drive
Little Rock, AR 72209
(501) 682-0833

**CALIFORNIA**
Department of Toxic Substances Control
P.O. Box 806
Sacramento, CA 95812-0806
(800) 728-6942

**COLORADO**
Hazardous Materials and Waste Management Division
Colorado Department of Public Health and Environment
Mail Code: HMWMD-HWC-B2
4300 Cherry Creek Drive, South
Denver, CO 80222-1530
(888) 569-1831

**CONNECTICUT**
Department of Environmental Protection
Waste Management Bureau
79 Elm Street
Hartford, CT 06106-5127
(860) 424-4081

# INFORMATIONAL RESOURCES *(cont.)*

**DELAWARE**
Department of Natural Resources
and Environmental Control
Division of Environmental Control
Solid Waste/Hazardous Waste Section
89 Kings Highway
Dover, DE 19901
(302) 739-3689

**DISTRICT OF COLUMBIA**
Hazardous Waste Environmental Health Administration
51 N. Street, 3rd Floor
Washington, DC 20002
(202) 535-2270

**FLORIDA**
Bureau of Solid and Hazardous Waste
Department of Environmental Protection
3900 Commonwealth Boulevard, M.S. 49
Tallahassee, FL 32399
(850) 245-8707

**GEORGIA**
Environmental Protection Division
Hazardous Waste Program
2 Martin Luther King Drive, SE
Atlanta, GA 30334
(404) 656-7802

**HAWAII**
State of Hawaii
Department of Health
Environmental Management Division

Solid & Hazardous Waste Branch
P.O. Box 3378
Honolulu, HI 96801-3378
(808) 586-4226

**IDAHO**

Department of Health and Welfare
Division of Environment
Bureau of Hazardous Materials
4040 Guard Street
Boise, ID 83705-5004
(208) 422-5725

**ILLINOIS**

State of Illinois
Environmental Protection Agency
Bureau of Land
1021 N Grand Avenue, East
Springfield, IL 62702
(217) 524-3300

**INDIANA**

Indiana Department of Environmental Management
Industrial Waste Permit Section
P.O. Box 6015
Indianapolis, IN 46206-6015
(317) 308-3003

**IOWA**

Department of Natural Resources
Solid Waste Section
Land Quality Bureau
502 East 9th Street

Des Moines, IA 50319-0034
(515) 281-8986

**KANSAS**
Department of Health and Environment
Bureau of Waste Management
1000 West Jackson Street, Suite 320
Topeka, KS 66612-1366
(785) 296-1601

**KENTUCKY**
NREPC
Division of Waste Management
Ft. Boone Plaza
14 Reilly Road
Frankfort, KY 40601
(502) 564-6716

**LOUISIANA**
Department of Environmental Quality
Office of Environmental Assessment
Solid Waste Division
P.O. Box 4314
Baton Rouge, LA 70821-4314
(225) 765-0355

**MAINE**
Department of Environmental Protection
Remediation and Waste Management
17 State House Station
Augusta, ME 04333-0017
(207) 287-7688

**MARYLAND**
Department of Environment
Hazardous Waste Program
1800 Washington Boulevard
Baltimore, MD 21230
(410) 537-3345

**MASSACHUSETTS**
Department of Environmental Protection
Office of Hazardous Waste
1 Winter Street
Boston, MA 02108
(617) 292-5898

**MICHIGAN**
Department of Natural Resources
Waste and Hazardous Materials Division
P.O. Box 30241
Lansing, MI 48909
(517) 373-2730

**MINNESOTA**
Minnesota Pollution Control Agency
Solid or Hazardous Waste Division
520 Lafayette Road North
St. Paul, MN 55155-4194
(651) 297-2274

**MISSISSIPPI**
Department of Environmental Quality
Office of Pollution Control
P.O. Box 20305
Jackson, MS 39289
(601) 961-5100

## INFORMATIONAL RESOURCES *(cont.)*

**MISSOURI**
Department of Natural Resources
Division of Environmental Quality
Hazardous Waste Program
Jefferson State Office Building
P.O. Box 176
Jefferson City, MO 65102
(573) 751-3176

**MONTANA**
Department of Environmental Quality
Permitting/Compliance Division
Air and Waste Management Division
P.O. Box 200901
Helena, MT 59620-0901
(406) 444-3490

**NEBRASKA**
Department of Environmental Quality
1200 N Street, Suite 400
P.O. Box 98922
State Office Building
Lincoln, NE 68509
(402) 471-2186

**NEVADA**
Bureau of Management Waste
Hazardous Waste Management Program
333 West Nye Lane
Carson City, NV 89706-0851
(775) 687-4670

**NEW HAMPSHIRE**
Department of Environmental Services
Waste Management Division/
Hazardous Waste Compliance
6 Hazen Drive
Concord, NH 03301
(603) 271-2942

**NEW JERSEY**
New Jersey Department of Environmental Protection
Division of Solid and Hazardous Waste
401 East State Street
P.O. Box 414
Trenton, NJ 08625-0414
(609) 633-1418

**NEW MEXICO**
New Mexico Environment Department
Hazardous Waste Bureau
2905 Rodeo Park Drive, East
Building 1
Santa Fe, NM 87505
(505) 428-2500

**NEW YORK**
Division of Solid and Hazardous Materials
New York State Department of
Environmental Conservation
625 Broadway
Albany, NY 12233
(518) 402-8712

**NORTH CAROLINA**
Department of Environment, Health,
and Natural Resources
Solid Waste Management/Hazardous Waste Division
1601 Mail Service Center
Raleigh, NC 27699-1601
(919) 733-4996

**NORTH DAKOTA**
Health Department, Environmental Health Section
Division of Waste Management
1200 Missouri Avenue
P.O. Box 5520
Bismarck, ND 58506-5520
(701) 328-5166

**OHIO**
Environmental Protection Agency
Office of Solid and Infectious Management
122 South Front Street
Columbus, OH 43215
(614) 644-3020

**OKLAHOMA**
Oklahoma Department of Environmental Quality
Land Protection Division
P.O. Box 1677
Oklahoma City, OK 73101-1677
(405) 702-5100

**OREGON**
Department of Environmental Quality
Waste Management Clean-up Division
811 Southwest 6th Avenue

Portland, OR 97204-1390
(503) 229-5913

**PENNSYLVANIA**
Department of Environmental Protection
Division of Hazardous Waste Management
P.O. Box 8471
Harrisburg, PA 17105-8471
(717) 787-6239

**PUERTO RICO**
Environmental Quality Board
Solid and Hazardous Waste Bureau
National Plaza Building, 12th Floor
Ponce de Leon Avenue, #431
Hato Rey, PR 00917
(787) 767-8056

**RHODE ISLAND**
Department of Environmental Management
Office of Waste Management
235 Promenade Street
Providence, RI 02908-5767
(401) 277-2797

**SOUTH CAROLINA**
Board of Health and Environmental Control
Bureau of Land and Waste Management
2600 Bull Street
Columbia, SC 29201
(803) 896-4000

**SOUTH DAKOTA**
Department of Water and Natural Resources
Environmental Health Division
Waste Management Program
Joe Foss Building
523 East Capitol
Pierre, SD 57501
(605) 773-3153

**TENNESSEE**
Department of Environment and Conservation
Division of Solid Waste Management
5th Floor, L&C Tower
401 Church Street
Nashville, TN 37243-1535
(615) 532-0780

**TEXAS**
Texas Commission on Environmental Quality
P.O. Box 13087
Austin, TX 78711-3087
(512) 239-2334

**UTAH**
Department of Environmental Quality
Division of Solid and Hazardous Waste
168 North 1950 West
Salt Lake City, UT 84116
(801) 538-6765

**VERMONT**
Department of Environmental Conservation
Environmental Assistance Division
103 South Main Street

Landry Building
Waterbury, VT 05671-0411
(802) 241-3589

**VIRGINIA**
Virginia Department of Environmental Quality
Waste Management
P.O. Box 10009
Richmond, VA 22240
(804) 698-4193

**WASHINGTON**
Department of Ecology
Solid and Hazardous Waste Program
P.O. Box 47600
Olympia, WA 98504-7600
(206) 407-6702

**WEST VIRGINIA**
West Virginia Division of Environmental Protection
Division of Water and Waste Management
1356 Hansford Street
Charleston, WV 25301
(304) 558-5989

**WISCONSIN**
Department of Natural Resources
Division of Air and Waste
P.O. Box 7921
101 South Webster Street
Madison, WI 53707-7921
(608) 266-2621

**WYOMING**

Department of Environmental Quality
Solid Waste Management Program
122 West 25th Street
Herscheler Building
Cheyenne, WY 82002
(307) 777-7752

**TSCA, RCRA and CERCLA**
**Information Phone Lines**

**Toxic Substances Control Act (TSCA)**
Assistance Information Hotline
(202) 554-1404

**RCRA/CERCLA Hotline**
(800) 424-9346
in the Washington, DC, Metro Area
(703) 412-9810

**CERCLA National Response Center Hotline**
(800) 424-8802

# CHAPTER 8
## *OSHA*

---

### OSHA

OSHA stands for the Occupational Safety and Health Administration, a branch of the U.S. Department of Labor, and ultimately of the U.S. federal government, originating with laws and mandates first passed in 1971. The general mission of OSHA is to keep workplaces safe for employees. As of 2005, OSHA had more than 2,220 employees, including 1,100 inspectors. The agency's budget was $468.1 million.

OSHA has had a variable history of being reasonable and otherwise. For example, at one point in the mid-1990s, OSHA began fining electrical contractors for not making their electricians wear rubber gloves and face masks while working on residential systems (this would have driven most electrical contractors out of business very quickly). After a big uproar they backed down, but serious fines were levied and businesses were hurt. Emphasis on fairness has returned, but what happens in the future is uncertain.

OSHA inspections (especially unannounced inspections) are of a disputed legality, but remain common. OSHA inspectors are authorized to issue violation notices, impose fines (which can be high) and set a time requirement for the correction of the violation(s). Employers are required to post these notices at the job site.

---

## OSHA *(cont.)*

Violation notices may be disputed for only 15 days. After that time, they are final. Any disputes regarding violations and fines go to an OSHA review commission. Employers unhappy with the commission's findings must take them to the U.S. Court of Appeals (a prohibitively expensive proposition).

Most OSHA departments are direct sections of the U.S. Department of Labor, but 20-some states have their own programs, which operate as partners with national OSHA.

It is important to understand that OSHA applies directly—and only—to *employers,* and NOT to *employees*. For example, an OSHA regulation that says, "unrolled wire mesh must be prevented from recoiling," seems to address the actions of a job site worker. After all, this is the person who will actually place ties on the mesh. But this requirement actually addresses the employer, who is required to make sure that the employee performs his or her work safely.

This application of rules on the employer has become a point of confusion as more post-traditional worker-employer relationships have entered the construction industry. So as long as you follow the OSHA rules all should be well, but do bear in mind that OSHA may consider you an employer, even if you are a single independent contractor.

| | **TOP 100 OSHA VIOLATIONS (2001)** | | |
|---|---|---|---|
| | **Section** | **Description** | **Total Violations** |
| 1 | .451(g)(1) | Scaffolding—Fall protection | 1846 |
| 2 | .501(b)(1) | Fall protection—Unprotected sides and edges | 1692 |
| 3 | .100(a) | PPE—Head protection | 1351 |
| 4 | .652(a)(1) | Excavations—Protection of employees in excavations | 1042 |
| 5 | .451(e)(1) | Scaffolding—Access | 1001 |
| 6 | .21(b)(2) | Safety training and education— Employer responsibility | 972 |
| 7 | .451(b)(1) | Scaffolding—Platform construction—Planking | 921 |
| 8 | .453(b)(2) | Scaffolding—Aerial lifts— Extensible and articulating boom platforms | 792 |
| 9 | .404(b)(1) | Electrical—Ground-fault protection | 758 |
| 10 | .451(c)(2) | Scaffolding—Supported scaffolds—Base plates | 756 |
| 11 | .503(a)(1) | Fall protection—Training program | 694 |
| 12 | .1053(b)(1) | Ladders—Use—3-foot extension above upper landing | 681 |
| 13 | .501(b)(13) | Fall protection—Residential construction | 679 |
| 14 | .20(b)(2) | Accident prevention responsibilities—Job site inspections | 653 |

|    | Section | Description | Total Violations |
|----|---------|-------------|------------------|
| 15 | 1910.1200(e)(1) | Hazard communication—Written hazard communication program | 638 |
| 16 | .454(a) | Scaffolding—Training requirements | 587 |
| 17 | .1052(c)(1) | Stairways—Stair rails and handrails | 579 |
| 18 | .20(b)(1) | Accident prevention responsibilities—Necessary programs | 558 |
| 19 | .501(b)(4) | Fall protection—Holes | 540 |
| 20 | .405(a)(2) | Electrical—Wiring methods—Temporary wiring | 495 |
| 21 | .651(k)(1) | Excavations—Inspections | 431 |
| 22 | .501(b)(10) | Fall protection—Roofing work on low-slope roofs | 430 |
| 23 | .451(g)(4) | Scaffolding—Guardrail systems | 425 |
| 24 | .651(c)(2) | Excavations—Access and egress | 424 |
| 25 | .102(a)(1) | PPE—Eye and face protection | 407 |
| 26 | .651(j)(2) | Excavations—Loose rock or soil | 377 |
| 27 | .25(a) | Housekeeping | 352 |
| 28 | .404(f)(6) | Electrical—Wiring design and protection—Grounding path | 346 |
| 29 | .701(b) | Concrete and masonry construction—Impalement hazards—Rebar | 340 |

| | | TOP 100 OSHA VIOLATIONS (2001) *(cont.)* | |
|---|---|---|---|
| | **Section** | **Description** | **Total Violations** |
| 30 | .451(f)(7) | Scaffolding—Erecting, dismantling, moving and altering | 326 |
| 31 | .405(g)(2) | Electrical—Identification, splices and terminations | 304 |
| 32 | .451(f)(3) | Scaffolding—Inspection | 302 |
| 33 | .501(b)(11) | Fall protection—Steep roofs | 271 |
| 34 | .95(a) | PPE—Criteria for personal protective equipment | 265 |
| 35 | 1910.1200(h)(1) | Hazard communication—Employee information and training | 255 |
| 36 | .1051(a) | Stairways and ladders—Personal points of access | 236 |
| 37 | .501(b)(14) | Fall protection—Wall openings | 233 |
| 38 | .403(b)(2) | Electrical—Examination, installation and use of equipment | 223 |
| 39 | 1910.1200(g)(1) | Hazard communication—Material safety data sheets | 222 |
| 40 | .150(c)(1) | Fire protection—Portable firefighting equipment | 207 |
| 41 | .452(c)(2) | Scaffolding—Fabricated frame scaffolds | 200 |
| 42 | .62(d)(2) | Lead—Protection of employees during exposure assessment | 195 |
| 43 | 1910.1200(h) | Hazard communication—Employee information and training | 179 |

| | Section | Description | Total Violations |
|---|---|---|---|
| 44 | .403(i)(2) | Electrical—Guarding electrical equipment live parts | 174 |
| 45 | .105(a) | PPE—Safety nets | 173 |
| 46 | .416(e)(1) | Electrical— Safety-related work practices—Cords and cables | 162 |
| 47 | .1060(a) | Stairways and ladders— Training requirements | 161 |
| 48 | .602(c)(1) | Material handling equipment— Lifting and hauling equipment | 157 |
| 49 | .454(b) | Scaffolding—Training requirements | 155 |
| 50 | .1053(b)(4) | Ladders—Use only for purpose designed for | 153 |
| 51 | .350(a)(9) | Gas welding and cutting— Gas cylinders | 152 |
| 52 | .451(h)(2) | Scaffolding—Falling object protection | 151 |
| 53 | 1910.1200(g)(8) | Hazard communication— Material safety data sheets | 150 |
| 54 | .1053(b)(13) | Ladders—Using top step | 141 |
| 55 | .405(b)(2) | Electrical—Wiring methods, components and equipment | 139 |
| 56 | .501(b)(15) | Fall protection—Walking/ working surfaces not otherwise addressed | 139 |
| 57 | .451(c)(1) | Scaffolding—Criteria for supported scaffolds | 130 |
| 58 | .405(b)(1) | Electrical—Wiring methods— Cabinets, boxes and fittings | 128 |

# TOP 100 OSHA VIOLATIONS (2001) *(cont.)*

| | Section | Description | Total Violations |
|---|---|---|---|
| 59 | .1053(b)(16) | Ladders—Structural defects | 126 |
| 60 | .403(b)(1) | Electrical—Approved equipment | 117 |
| 61 | 1910.134(c)(1) | Respiratory protection—Respiratory protection program | 116 |
| 62 | .350(a)(10) | Welding and cutting—Storing gas cylinders | 116 |
| 63 | .602(a)(9) | Material handling equipment—Earth-moving equipment | 113 |
| 64 | .451(b)(2) | Scaffolding—Platform construction | 112 |
| 65 | 1910.134(d)(1) | Respiratory protection—Selection of respirators | 110 |
| 66 | .62(d)(1) | Lead—Exposure assessment | 110 |
| 67 | 1910.178(1) | Powered industrial trucks (forklifts)—Operator training | 109 |
| 68 | .651(k)(2) | Excavations—Inspections | 109 |
| 69 | .501(b)(3) | Fall protection—Hoist areas | 107 |
| 70 | 1910.134(e)(1) | Respiratory protection—Medical evaluation | 104 |
| 71 | .304(f) | Tools—Woodworking tools | 104 |
| 72 | .451(b)(5) | Scaffolding—Platform construction | 102 |
| 73 | .502(b)(2) | Fall protection systems—Guardrail systems | 99 |
| 74 | .28(a) | Personal protective equipment—Employer responsibility | 97 |

| | Section | Description | Total Violations |
|---|---|---|---|
| 75 | .503(c) | Fall protection—Retraining | 97 |
| 76 | .503(b)(1) | Fall protection—Certification of training | 92 |
| 77 | .1052(c)(12) | Stairways—Guardrails | 91 |
| 78 | .451(h)(1) | Scaffolding—Falling object protection | 89 |
| 79 | .502(b)(1) | Fall protection systems—Guardrail systems | 89 |
| 80 | .1052(b)(1) | Stairways—Temporary service | 87 |
| 81 | .451(a)(6) | Scaffolding—Capacity | 86 |
| 82 | .452(w)(2) | Scaffolding—Mobile scaffolds | 86 |
| 83 | .502(i)(4) | Fall protection—Covers | 84 |
| 84 | .503(a)(2) | Fall protection—Training program | 83 |
| 85 | .503(c)(3) | Fall protection—Retraining | 81 |
| 86 | .416(a)(1) | Electrical—Protection of employees | 80 |
| 87 | .1053(b)(8) | Ladders—Securing ladders | 80 |
| 88 | .405(j)(1) | Electrical—Lighting fixtures, lamps and receptacles | 78 |
| 89 | .502(i)(3) | Fall protection—Covers | 77 |
| 90 | .451(f)(4) | Scaffolding—Damaged or weakened | 75 |
| 91 | .651(d) | Excavations—Exposure to vehicular traffic | 74 |
| 92 | 1910.134(f)(2) | Respiratory protection—Fit testing | 72 |

## TOP 100 OSHA VIOLATIONS (2001) *(cont.)*

| | Section | Description | Total Violations |
|---|---|---|---|
| 93 | .200(g)(1) | Accident prevention signs and tags—Traffic signs | 70 |
| 94 | .405(g)(1) | Electrical—Use of flexible cords and cables | 69 |
| 95 | .451(b)(4) | Scaffolding—Platform construction | 67 |
| 96 | .1053(b)(6) | Ladders—Use | 67 |
| 97 | .152(a)(1) | Flammable and combustible liquids—General requirements | 62 |
| 98 | .300(b)(1) | Tools, hand and power—Guarding | 62 |
| 99 | .300(b)(2) | Tools, hand and power—Guarding | 60 |
| 100 | .502(b)(3) | Fall protection—Guardrail systems | 60 |

## OSHA SAFETY COLOR CODES

| Color | Examples |
|---|---|
| Red | Fire protection equipment and apparatus; portable containers of flammable liquids; emergency stop pushbuttons/switches |
| Yellow | Caution and for marking physical hazards, waste containers for explosive or combustible materials; caution against starting, using or moving equipment under repair; identification of the starting point or power source of machinery |
| Orange | Dangerous parts of machines; safety starter buttons; the exposed parts of pulleys, gears, rollers, cutting devices and power jaws |
| Purple | Radiation hazards |
| Green | Safety areas and location of first aid equipment |

## HAZARD ASSESSMENT *(cont.)*

include a review of injury and illness records to spot any trends or areas of concern and appropriate corrective action. The suitability of existing PPE, including an evaluation of its condition and age, should be included in the reassessment.

Documentation of the hazard assessment is required through a written certification that includes the following information:

- Identification of the workplace evaluated;
- Name of the person conducting the assessment;
- Date of the assessment; and
- Identification of the document certifying completion of the hazard assessment.

## SELECTING PERSONAL PROTECTIVE EQUIPMENT

All PPE clothing and equipment should be of safe design and construction, and it should be maintained in a clean and reliable fashion. Employers should take the fit and comfort of PPE into consideration when selecting appropriate items for their workplace. PPE that fits well and is comfortable to wear will encourage employee use. Most protective devices are available in multiple sizes, and care should be taken to select the proper size for each employee. If several types of PPE are worn together, make sure they are compatible. If PPE does not fit properly, it can make the difference between being safely covered or dangerously exposed. It may not provide the level of protection desired and may discourage employee use.

## SELECTING PERSONAL PROTECTIVE EQUIPMENT *(cont.)*

Employers who need to provide PPE in the categories listed below must make certain that any new equipment procured meets the cited American National Standards Institute (ANSI) standard.

Employers should inform employees who provide their own PPE of the employer's selection decisions and ensure that any employee-owned PPE used in the workplace conforms to the employer's criteria, based on the hazard assessment, OSHA requirements and ANSI standards. OSHA requires PPE to meet the following ANSI standards:

- Eye and Face Protection: ANSI Z87.1-1989 (USA Standard for Occupational and Educational Eye and Face Protection).
- Head Protection: ANSI Z89.1-1986.
- Foot Protection: ANSI Z41.1-1991.

For hand protection, there is no ANSI standard, but OSHA recommends that selection be based on the tasks to be performed and the performance and construction characteristics of the glove material. For protection against chemicals, glove selection must be based on the chemicals encountered, the chemical resistance and the physical properties of the glove material.

# TRAINING EMPLOYEES IN THE PROPER USE OF PPE

Employers are required to train each employee who must use PPE. Employees must be trained to know at least the following:

- When PPE is necessary.
- What PPE is necessary.
- How to properly put on, take off, adjust and wear the PPE.
- The limitations of the PPE.
- Proper care, maintenance, useful life and disposal of PPE.

Employers should make sure that each employee demonstrates an understanding of the PPE training as well as the ability to properly wear and use PPE before they are allowed to perform work requiring the use of the PPE. If an employer believes that a previously trained employee is not demonstrating the proper understanding and skill level in the use of PPE, that employee should receive retraining. Other situations that require additional or retraining of employees include changes in the workplace or in the type of required PPE that make prior training obsolete.

The employer must document the training of each employee required to wear or use PPE by preparing a certificate containing the name of each employee trained, the date of training and a clear identification of the subject of the certification.

## EYE AND FACE PROTECTION

OSHA requires employers to ensure that employees have appropriate eye or face protection if they are exposed to eye or face hazards from flying particles, molten metal, liquid chemicals, acids or caustic liquids, chemical gases or vapors, potentially infected material or potentially harmful light radiation.

Employers must be sure that their employees wear appropriate eye and face protection and that the selected form of protection is appropriate to the work being performed and properly fits each worker exposed to the hazard.

Everyday use of prescription corrective lenses will not provide adequate protection against most occupational eye and face hazards, so employers must make sure that employees with corrective lenses either wear eye protection that incorporates the prescription into the design or wear additional eye protection over their prescription lenses. Examples of potential eye or face injuries include:

- Dust, dirt, metal or wood chips entering the eye from activities such as chipping, grinding, sawing, hammering, the use of power tools or even strong wind forces.
- Chemical splashes from corrosive substances, hot liquids, solvents or other hazardous solutions.
- Objects swinging into the eye or face, such as tree limbs, chains, tools or ropes.
- Radiant energy from welding, harmful rays from the use of lasers or other radiant light (as well as heat, glare, sparks, splash and flying particles).

## EYE PROTECTION

Selecting the most suitable eye and face protection for employees should take into consideration the following elements:

- Ability to protect against specific workplace hazards.
- Should fit properly and be reasonably comfortable to wear.
- Should provide unrestricted vision and movement.
- Should be durable and cleanable.
- Should allow unrestricted functioning of any other required PPE.

An employer may choose to provide one pair of protective eyewear for each position rather than individual eyewear for each employee.

Some of the most common types of eye and face protection include the following:

- **Safety spectacles.** These protective eyeglasses have safety frames constructed of metal or plastic and impact-resistant lenses. Side shields are available on some models.
- **Goggles.** These are tight-fitting eye protection that completely cover the eyes, eye sockets and the facial area immediately surrounding the eyes and provide protection from impact, dust and splashes. Some goggles will fit over corrective lenses.
- **Welding shields.** Constructed of vulcanized fiber or fiberglass and fitted with a filtered lens; welding shields protect eyes from burns caused by infrared or intense radiant light; they also

protect both the eyes and face from flying sparks, metal spatter and slag chips produced during welding, brazing, soldering and cutting operations. OSHA requires filter lenses to have a shade number appropriate to protect against the specific hazards of the work being performed in order to protect against harmful light radiation.

- **Laser safety goggles.** These specialty goggles protect against intense concentrations of light produced by lasers. The type of laser safety goggles an employer chooses will depend on the equipment and operating conditions in the workplace.

- **Face shields.** These transparent sheets of plastic extend from the eyebrows to below the chin and across the entire width of the employee's head. Some are polarized for glare protection. Face shields protect against nuisance dusts and potential splashes or sprays of hazardous liquids but will not provide adequate protection against impact hazards. Face shields used in combination with goggles or safety spectacles will provide additional protection against impact hazards.

## HEAD PROTECTION

Employers must ensure that their employees wear head protection if any of the following apply:

- Objects might fall from above and strike them on the head;
- They might bump their heads against fixed objects, such as exposed pipes or beams; or

- There is a possibility of accidental head contact with electrical hazards.

Examples of occupations in which employees should be required to wear head protection include construction workers, carpenters, electricians, linemen, plumbers and pipe fitters, timber and log cutters and welders. Whenever there is a danger of objects falling from above, such as working below others who are using tools or working under a conveyor belt, head protection must be worn. Hard hats must be worn with the bill forward to protect employees properly.

In general, protective helmets or hard hats should do the following:

- Resist penetration by objects.
- Absorb the shock of a blow.
- Be water-resistant and slow burning.
- Have clear instructions explaining proper adjustment and replacement of the suspension and headband.

Hard hats must have a hard outer shell and a shock-absorbing lining that incorporates a headband and straps that suspend the shell from 1 to 1¼" (2.54 cm to 3.18 cm) away from the head. This type of design provides shock absorption during an impact and ventilation during normal wear. Hard hats are divided into three industrial classes:

- **Class A** hard hats provide impact and penetration resistance along with limited voltage protection (up to 2,200 volts).

## HEAD PROTECTION (cont.)

- **Class B** hard hats provide the highest level of protection against electrical hazards, with high-voltage shock and burn protection (up to 20,000 volts). They also provide protection from impact and penetration hazards by flying/falling objects.
- **Class C** hard hats provide lightweight comfort and impact protection but offer no protection from electrical hazards.

Hard hats with any of the following defects should be removed from service and replaced:

- Perforation, cracking or deformity of the brim or shell.
- Indication of exposure of the brim or shell to heat, chemicals or ultraviolet light and other radiation (in addition to a loss of surface gloss, such signs include chalking or flaking).

## FOOT AND LEG PROTECTION

Situations in which an employee should wear foot and/or leg protection include:

- When heavy objects such as barrels or tools might roll or fall onto the employee's feet;
- Working with sharp objects such as nails or spikes that could pierce the soles or uppers of ordinary shoes;
- Exposure to molten metal that might splash on feet or legs;

- Working on or around hot, wet or slippery surfaces; and
- Working when electrical hazards are present.

Foot and leg protection choices include the following:

- **Leggings** protect the lower legs and feet from heat hazards such as molten metal or welding sparks. Safety snaps allow leggings to be removed quickly.
- **Metatarsal guards** protect the instep area from impact and compression. Made of aluminum, steel, fiber or plastic, these guards may be strapped to the outside of shoes.
- **Toe guards** fit over the toes of regular shoes to protect from impact and compression hazards. They may be made of steel, aluminum or plastic.
- **Combination foot and shin guards** protect the lower legs and feet, and may be used in combination with toe guards when greater protection is needed.
- **Safety shoes** have impact-resistant toes and heat-resistant soles that protect the feet against hot work surfaces common in roofing, paving and hot metal industries.

### Special Purpose Shoes

**Electrically conductive shoes** provide protection against the buildup of static electricity. Employees working in explosive and hazardous locations such as explosives manufacturing facilities or grain elevators must wear conductive shoes to reduce the risk of static electricity buildup on the body that could produce a spark and cause an explosion or fire. Foot powder should not be used in conjunction with protective conductive footwear because it provides insulation, reducing the conductive ability of the shoes. Silk, wool and nylon socks can produce static electricity and should not be worn with conductive footwear. Conductive shoes must be removed when the task requiring their use is completed. Employees exposed to electrical hazards must never wear conductive shoes.

**Electrical hazard, safety-toe shoes** are nonconductive and will prevent the wearers' feet from completing an electrical circuit to the ground. These shoes can protect against open circuits of up to 600 volts in dry conditions and should be used in conjunction with other insulating equipment and additional precautions to reduce the risk of a worker becoming a path for hazardous electrical energy. The insulating protection of electrical hazard, safety-toe shoes may be compromised if the shoes become wet, the soles are worn through, metal particles become embedded in the sole or heel or workers touch conductive, grounded items. Nonconductive footwear must not be used in explosive or hazardous locations.

### Foundry Shoes

In addition to insulating the feet from the extreme heat of molten metal, foundry shoes keep hot metal from lodging in shoe eyelets, tongues or other shoe parts. These snug-fitting leather or leather-substitute shoes have leather or rubber soles and rubber heels. All foundry shoes must have built-in safety toes.

### Care of Protective Footwear

As with all protective equipment, safety footwear should be inspected prior to each use. Shoes and leggings should be checked for wear and tear at reasonable intervals. This includes looking for cracks or holes, separation of materials, broken buckles or laces. The soles of shoes should be checked for pieces of metal or other embedded items that could present electrical or tripping hazards. Employees should follow the manufacturer's recommendations for cleaning and maintenance of protective footwear.

## HAND AND ARM PROTECTION

If a workplace hazard assessment reveals that employees face potential injury to hands and arms that cannot be eliminated through engineering and work practice controls, employers must ensure that employees wear appropriate protection. Potential hazards include skin absorption of harmful substances, chemical or thermal burns, electrical dangers, bruises, abrasions, cuts, punctures, fractures and amputations. Protective equipment includes gloves, finger guards and arm coverings or elbow-length gloves.

## HAND AND ARM PROTECTION *(cont.)*

The following are examples of some factors that may influence the selection of protective gloves for a workplace.

- Type of chemicals handled.
- Nature of contact (total immersion, splash, etc.).
- Duration of contact.
- Area requiring protection (hand only, forearm, arm).
- Grip requirements (dry, wet, oily).
- Thermal protection.
- Size and comfort.
- Abrasion/resistance requirements.
- Materials. Gloves are designed for many types of workplace hazards. In general, they fall into four groups:
  - Gloves made of leather, canvas or metal mesh;
  - Fabric and coated fabric gloves;
  - Chemical- and liquid-resistant gloves; and
  - Insulating rubber gloves.

### Glove Types

- **Leather gloves** protect against sparks, moderate heat, blows, chips and rough objects.
- **Aluminized gloves** provide reflective and insulating protection against heat and require an insert made of synthetic materials to protect against heat and cold.
- **Aramid fiber gloves** protect against heat and cold, are cut- and abrasive-resistant and wear well.

- **Synthetic gloves** of various materials offer protection against heat and cold, are cut- and abrasive-resistant and may withstand some diluted acids. These materials do not stand up against alkalis and solvents.

- **Fabric gloves** protect against dirt, slivers, chafing and abrasions. They do not provide sufficient protection for use with rough, sharp or heavy materials. Adding a plastic coating will strengthen some fabric gloves.

- **Coated fabric gloves** are normally made from cotton flannel with napping on one side. By coating the unnapped side with plastic, fabric gloves are transformed into general-purpose hand protection offering slip-resistant qualities. These gloves are used for tasks ranging from handling bricks and wire to chemical laboratory containers.

### Chemical-resistant gloves

- **Butyl gloves** are made of a synthetic rubber and protect against a wide variety of chemicals, such as peroxide, rocket fuels, highly corrosive acids (e.g., nitric acid, sulfuric acid, hydrofluoric acid and red-fuming nitric acid), strong bases, alcohols, aldehydes, ketones, esters and nitro-compounds.

- **Natural (latex) rubber gloves** are comfortable to wear, which makes them a popular general-purpose glove. They feature outstanding tensile strength, elasticity and temperature resistance.

In addition to resisting abrasions caused by grinding and polishing, these gloves protect workers' hands from most water solutions of acids, alkalis, salts and ketones.

- **Neoprene gloves** are made of synthetic rubber and offer good pliability, finger dexterity, high density and tear resistance. They protect against hydraulic fluids, gasoline, alcohols, organic acids and alkalis. They generally have chemical and wear resistance properties superior to those made of natural rubber.

- **Nitrile gloves** are made of a copolymer and provide protection from chlorinated solvents such as trichloroethylene and perchloroethylene. Although intended for jobs requiring dexterity and sensitivity, nitrile gloves stand up to heavy use even after prolonged exposure to substances that cause other gloves to deteriorate. They offer protection when working with oils, greases, acids, caustics and alcohols but are generally not recommended for use with strong oxidizing agents, aromatic solvents, ketones and acetates.

## CHEMICAL RESISTANCE SELECTION CHART FOR PROTECTIVE GLOVES

| Chemical | Neoprene | Latex/Rubber | Butyl | Nitrile |
|---|---|---|---|---|
| Acetaldehyde* | VG | G | VG | G |
| Acetic acid | VG | VG | VG | VG |
| Acetone* | G | VG | VG | P |
| Ammonium hydroxide | VG | VG | VG | VG |
| Amy acetate* | F | P | F | P |
| Aniline | G | F | F | P |
| Benzaldehyde* | F | F | G | G |
| Benzene* | P | P | P | F |
| Butyl acetate | G | F | F | P |
| Butyl alcohol | VG | VG | VG | VG |
| Carbon disulfide | F | F | F | F |
| Carbon tetrachloride* | F | P | P | G |
| Castor oil | F | P | F | VG |
| Chlorobenzene* | F | P | F | P |
| Chloroform* | G | P | P | F |
| Chloronaphthalene | F | P | F | F |
| Chromic acid (50%) | F | P | F | F |
| Citric acid (10%) | VG | VG | VG | VG |
| Cyclohexanol | G | F | G | VG |
| Dibutyl phthalate* | G | P | G | G |
| Diesel fuel | G | P | P | VG |
| Diisobutyl ketone | P | F | G | P |
| Dimethylformamide | F | F | G | G |
| Dioctyl phthalate | G | P | F | VG |
| Dioxane | VG | G | G | G |

VG: Very Good; G: Good; F: Fair; P: Poor (not recommended).
Chemicals marked with an asterisk (*) are for limited service.

## CHEMICAL RESISTANCE SELECTION CHART FOR PROTECTIVE GLOVES (cont.)

| Chemical | Neoprene | Latex/Rubber | Butyl | Nitrile |
|---|---|---|---|---|
| Epoxy resins, dry | VG | VG | VG | VG |
| Ethyl acetate* | G | F | G | F |
| Ethyl alcohol | VG | VG | VG | VG |
| Ethyl ether* | VG | G | VG | G |
| Ethylene dichloride* | F | P | F | P |
| Ethylene glycol | VG | VG | VG | VG |
| Formaldehyde | VG | VG | VG | VG |
| Formic acid | VG | VG | VG | VG |
| Freon 11 | G | P | F | G |
| Freon 12 | G | P | F | G |
| Freon 21 | G | P | F | G |
| Freon 22 | G | P | F | G |
| Furfural* | G | G | G | G |
| Gasoline, leaded | G | P | F | VG |
| Gasoline, unleaded | G | P | F | VG |
| Glycerin | VG | VG | VG | VG |
| Hexane | F | P | P | G |
| Hydrazine (65%) | F | G | G | G |
| Hydrochloric acid | VG | G | G | G |
| Hydrofluoric acid (48%) | VG | G | G | G |
| Hydrogen peroxide (30%) | G | G | G | G |
| Hydroquinone | G | G | G | F |
| Isooctane | F | P | P | VG |
| Kerosene | VG | F | F | VG |
| Ketones | G | VG | VG | P |

VG: Very Good; G: Good; F: Fair; P: Poor (not recommended).
Chemicals marked with an asterisk (*) are for limited service.

# CHEMICAL RESISTANCE SELECTION CHART FOR PROTECTIVE GLOVES (cont.)

| Chemical | Neoprene | Latex/Rubber | Butyl | Nitrile |
|---|---|---|---|---|
| Lacquer thinners | G | F | F | P |
| Lactic acid (85%) | VG | VG | VG | VG |
| Lauric acid (36%) | VG | F | VG | VG |
| Lineolic acid | VG | P | F | G |
| Linseed oil | VG | P | F | VG |
| Maleic acid | VG | VG | VG | VG |
| Methyl alcohol | VG | VG | VG | VG |
| Methylamine | F | F | G | G |
| Methyl bromide | G | F | G | F |
| Methyl chloride* | P | P | P | P |
| Methyl ethyl ketone* | G | G | VG | P |
| Methyl isobutyl ketone* | F | F | VG | P |
| Methyl metharcrylate | G | G | VG | F |
| Monoethanolamine | VG | G | VG | VG |
| Morpholine | VG | VG | VG | G |
| Naphthalene | G | F | F | G |
| Napthas, aliphatic | VG | F | F | VG |
| Napthas, aromatic | G | P | P | G |
| Nitric acid* | G | F | F | F |
| Nitric acid, red and white fuming | P | P | P | P |
| Nitromethane (95.5%)* | F | P | F | F |
| Nitropropane (95.5%) | F | P | F | F |
| Octyl alcohol | VG | VG | VG | VG |
| Oleic acid | VG | F | G | VG |

VG: Very Good; G: Good; F: Fair; P: Poor (not recommended).
Chemicals marked with an asterisk (*) are for limited service.

## CHEMICAL RESISTANCE SELECTION CHART FOR PROTECTIVE GLOVES *(cont.)*

| Chemical | Neoprene | Latex/Rubber | Butyl | Nitrile |
|---|---|---|---|---|
| Oxalic acid | VG | VG | VG | VG |
| Palmitic acid | VG | VG | VG | VG |
| Perchloric acid (60%) | VG | F | G | G |
| Perchloroethylene | F | P | P | G |
| Petroleum distillates (naphtha) | G | P | P | VG |
| Phenol | VG | F | G | F |
| Phosphoric acid | VG | G | VG | VG |
| Potassium hydroxide | VG | VG | VG | VG |
| Propyl acetate | G | F | G | F |
| Propyl alcohol | VG | VG | VG | VG |
| Propyl alcohol (iso) | VG | VG | VG | VG |
| Sodium hydroxide | VG | VG | VG | VG |
| Styrene | P | P | P | F |
| Styrene (100%) | P | P | P | F |
| Sulfuric acid | G | G | G | G |
| Tannic acid (65) | VG | VG | VG | VG |
| Tetrahydrofuran | P | F | F | F |
| Toluene* | F | P | P | F |
| Toluene diisocyanate (TDI) | F | G | G | F |
| Trichloroethylene* | F | F | P | G |
| Triethanolamine (85%) | VG | G | G | VG |
| Tung oil | VG | P | F | VG |
| Turpentine | G | F | F | VG |
| Xylene* | P | P | P | F |

VG: Very Good; G: Good; F: Fair; P: Poor (not recommended). Chemicals marked with an asterisk (*) are for limited service.

## ELECTRICAL GLOVE INSPECTION

- Gloves shall be inspected before and after every use.

- Check that the voltage rating of the insulating equipment is greater than the application voltage.

- Check for defects by stretching a small area at a time.

- Look for cracks, cuts, scratches, embedded material or weak spots.

- Inspect the entire glove. Any detectable flaw is cause for gloves to be returned for testing.

- Check for pinhole leaks by trapping air in the fingers of the glove. Also squeeze the cuff with one hand and inspect the palm, thumb and fingers for defects.

- Hold the glove to your face to check for air leakage.

- Any leakage is cause for the glove to be destroyed. Cut the glove in half immediately.

- Check the test date stamped (month and year) on any rubber protective equipment.

- Maximum time between tests shall be 6 months for gloves and 12 months for sleeves and insulating blankets.

## CARE OF PROTECTIVE GLOVES

Protective gloves should be inspected before each use to ensure that they are not torn, punctured or made ineffective in any way. A visual inspection will help detect cuts or tears, but a more thorough inspection by filling the gloves with water and tightly rolling the cuff toward the fingers will help reveal any pinhole leaks. Gloves that are discolored or stiff may also indicate deficiencies caused by excessive use or degradation from chemical exposure.

Any gloves with impaired protective ability should be discarded and replaced. Reuse of chemical-resistant gloves should be evaluated carefully, taking into consideration the absorptive qualities of the gloves. A decision to reuse chemically exposed gloves should take into consideration the toxicity of the chemicals involved and factors such as duration of exposure, storage and temperature.

## BODY PROTECTION

The following are examples of workplace hazards that could cause bodily injury:

- Temperature extremes.
- Hot splashes from molten metals and other hot liquids.
- Potential impacts from tools, machinery and materials.
- Hazardous chemicals.

If a hazard assessment indicates a need for full body protection against toxic substances or harmful physical agents, the clothing should be carefully inspected before each use, must fit each worker properly and must function properly and for the purpose for which it is intended.

### Protective Clothing Materials

- **Paper-like fiber** used for disposable suits provide protection against dust and splashes.
- **Treated wool and cotton** adapt well to changing temperatures, are comfortable and fire-resistant and protect against dust, abrasions and rough and irritating surfaces.
- **Duck** is a closely woven cotton fabric that protects against cuts and bruises when handling heavy, sharp or rough materials.
- **Leather** is often used to protect against dry heat and flames.
- **Rubber, rubberized fabrics, neoprene and plastics** protect against certain chemicals and physical hazards. When chemical or physical hazards are present, check with the clothing manufacturer to ensure that the material selected will provide protection against the specific hazard.

## HEARING PROTECTION

Employee exposure to excessive noise depends on a number of factors, including:

- The loudness of the noise as measured in decibels (dB).
- The duration of each employee's exposure to the noise.

- Whether employees move between work areas with different noise levels.
- Whether noise is generated from one or multiple sources.

Generally, the louder the noise, the shorter the exposure time before hearing protection is required. For instance, employees may be exposed to a noise level of 90 dB for 8 hours per day (unless they experience a Standard Threshold Shift) before hearing protection is required. On the other hand, if the noise level reaches 115 dB, hearing protection is required if the anticipated exposure exceeds 15 minutes.

Noises are considered continuous if the interval between occurrences of the maximum noise level is 1 second or less. Noises not meeting this definition are considered impact or impulse noises (loud momentary explosions of sound), and exposures to this type of noise must not exceed 140 dB. Examples of situations or tools that may result in impact or impulse noises are powder-actuated nail guns, a punch press or drop hammers.

**Types of Hearing Protection**

- **Single-use earplugs** are made of waxed cotton, foam, silicone rubber or fiberglass wool. They are self-forming and, when properly inserted, they work as well as most molded earplugs.
- **Pre-formed or molded earplugs** must be individually fitted by a professional and can be

disposable or reusable. Reusable plugs should be cleaned after each use.

- **Earmuffs** require a perfect seal around the ear. Glasses, facial hair, long hair or facial movements such as chewing may reduce the protective value of earmuffs.

## JACKS

All jacks—including lever and ratchet jacks, screw jacks, and hydraulic jacks—must have a stop indicator, and the stop limit must not be exceeded. The manufacturer's load limits must not be exceeded.

A jack should never be used to support a lifted load. Once the load has been lifted, it must immediately be blocked up. Put a block under the base of the jack when the foundation is not firm and place a block between the jack cap and load if the cap might slip. To set up a jack, make certain of the following:

- The base of the jack rests on a firm, level surface;
- The jack is correctly centered;
- The jack head bears against a level surface; and
- The lift force is applied evenly.

Proper maintenance of jacks is essential for safety.

All jacks must be lubricated regularly. In addition, each jack must be inspected according to the following schedule:

- Jacks used continuously or intermittently at one site: at least once every 6 months.
- Jacks sent out of the shop for special work: when sent out and when returned.
- Jacks subjected to abnormal loads or shock: before and immediately after use.

## CONCRETE

### Formwork

Formwork must be designed, fabricated, erected, supported, braced and maintained so that it will be capable of supporting without failure all vertical and lateral loads that might be applied to the formwork.

Drawings and plans, including all revisions for the jack layout, formwork (including shoring equipment), working decks and scaffolds, must be available at the jobsite.

### Shoring and Reshoring

All shoring equipment (including equipment used in reshoring operations) must be inspected prior to erection to determine whether the equipment meets the requirements specified in the formwork drawings.

Damaged shoring equipment must not be used for shoring. Erected shoring equipment must be inspected immediately prior to, during and immediately after concrete placement. Shoring equipment that is found to be damaged or weakened after erection must be immediately reinforced.

The sills for shoring must be sound, rigid and capable of carrying the maximum intended load. All base plates, shore heads, extension devices and adjustment screws must be in firm contact and secured, when necessary, with the foundation and the form.

Eccentric loads on shore heads must be prohibited unless these members have been designed for such loading.

If single-post shores are used one on top of another (tiered), then additional shoring requirements must be met. The shores must be as follows:

- Designed by a qualified designer, and the erected shoring must be inspected by an engineer qualified in structural design;
- Vertically aligned;
- Spliced to prevent misalignment; and
- Adequately braced in two mutually perpendicular directions at the splice level. Each tier also must be diagonally braced in the same two directions.

Adjustment of single-post shores to raise formwork must not be made after the placement of concrete.

Reshoring must be erected, as the original forms and shores are removed, whenever the concrete is required to support loads in excess of its capacity.

### Vertical Slip Forms

The steel rods or pipes on which jacks climb or by which the forms are lifted must be (1) specifically designed for that purpose and (2) adequately braced where not encased in concrete. Forms must be designed to prevent excessive distortion of the structure during the jacking operation. Jacks and vertical supports must be positioned in such a manner that the loads do not exceed the rated capacity of the jacks.

The jacks or other lifting devices must be provided with mechanical dogs or other automatic holding devices to support the slip forms whenever failure of the power supply or lifting mechanisms occurs.

The form structure must be maintained within all design tolerances specified for plumbness during the jacking operation.

Do not exceed the predetermined safe rate of the lift.

All vertical slip forms must be provided with scaffolds or work platforms where employees are required to work or pass.

### Reinforcing Steel

Reinforcing steel for walls, piers, columns and similar vertical structures must be adequately supported to prevent overturning and collapse.

Employers must take measures to prevent unrolled wire mesh from recoiling. Such measures may include, but are not limited to, securing each end of the roll or turning over the roll.

### Removal of Formwork

Forms and shores (except those that are used for slabs on grade and slip forms) must not be removed until the employer determines that the concrete has gained sufficient strength to support its weight and superimposed loads. Such determination must be based on compliance with one of the following:

- The plans and specifications stipulate conditions for removal of forms and shores, and such conditions have been followed, or
- The concrete has been properly tested.

Reshoring must not be removed until the concrete being supported has attained adequate strength to support its weight and all loads placed on it.

### Precast Concrete

Precast concrete wall units, structural framing and tilt-up wall panels must be adequately supported to prevent overturning and to prevent collapse until permanent connections are completed.

Lifting inserts that are embedded or otherwise attached to tilt-up wall panels must be capable of supporting at least two times the maximum intended load applied or transmitted to them; lifting inserts for other precast members must be capable of supporting four times the load. Lifting hardware shall be capable of supporting at least five times the maximum intended load applied or transmitted to the lifting hardware.

Only essential employees are permitted under precast concrete that is being lifted or tilted into position.

### Lift-Slab Operations

- Lift-slab operations must be designed and planned by a registered professional engineer who has experience in lift-slab construction.

- Jacking equipment must be marked with the manufacturer's rated capacity and must be capable of supporting at least two and one-half times the load being lifted during jacking operations, and the equipment must not be overloaded. Jacking equipment includes any load-bearing component used to carry out the lifting operation(s). Such equipment includes, but is not limited to, threaded rods, lifting attachments, lifting nuts, hook-up collars, T-caps, shearheads, columns and footings.

- Jacks/lifting units must be designed and installed so that they will neither lift nor continue to lift when loaded in excess of their rated capacity; and jacks/lifting units must have a safety device that will cause the jacks/lifting units to support the load at any position in the event of their malfunction or loss of ability to continue to lift.

- No employee, except those essential to the jacking operation, shall be permitted in the building/structure while any jacking operation is taking place unless the building/structure has been reinforced sufficiently to ensure its integrity during erection.

- Under no circumstances shall any employee who is not essential to the jacking operation be permitted immediately beneath a slab while it is being lifted.

## Masonry Construction

Whenever a masonry wall is being constructed, employers must establish a limited access zone prior to the start of construction. The limited access zone must be as follows:

- Equal to the height of the wall to be constructed plus 4' (1.2 m), and shall run the entire length of the wall;
- On the side of the wall that will be unscaffolded;
- Restricted to entry only by employees actively engaged in constructing the wall; and
- Kept in place until the wall is adequately supported to prevent overturning and collapse unless the height of the wall is more than 8' (2.4 m) and unsupported, in which case it must be braced. The bracing must remain in place until permanent supporting elements of the structure are in place.

## Reinforcing Steel

All protruding reinforcing steel, onto and into which employees could fall, must be guarded to eliminate the hazard of impalement.

## Post-Tensioning Operations

Employees (except those essential to the post-tensioning operations) must not be permitted to be behind the jack during tensioning operations.

Signs and barriers must be erected to limit employee access to the post-tensioning area during tensioning operations.

### Concrete Buckets

Employees must not be permitted to ride concrete buckets.

### Working Under Loads

Employees must not be permitted to work under concrete buckets while the buckets are being elevated or lowered into position.

To the extent practicable, elevated concrete buckets must be routed so that no employee or the fewest employees possible are exposed to the hazards associated with falling concrete buckets.

### Personal Protective Equipment

Employees must not be permitted to apply a cement, sand and water mixture through a pneumatic hose unless they are wearing protective head and face equipment.

### Equipment and Tools

The standard also includes requirements for the following equipment and operations:

- Bulk cement storage,
- Concrete mixers,
- Power concrete trowels,
- Concrete buggies,
- Concrete pumping systems,
- Concrete buckets,
- Tremies,
- Bull floats,
- Masonry saws, and
- Lockout/tagout procedures.

## SAFETY NETS

Safety nets must be installed as close as practicable under the walking/working surface on which employees are working and never more than 30' (9.1 m) below such levels. Defective nets shall not be used. Safety nets shall be inspected at least once a week for wear, damage and other deterioration. The maximum size of each safety net mesh opening shall not exceed 36 square inches (230 cm$^2$) nor be longer than 6" (15 cm) on any side, and the openings measured center-to-center, of mesh ropes or webbing, shall not exceed 6" (15 cm). All mesh crossings shall be secured to prevent enlargement of the mesh opening. Each safety net or section shall have a border rope for webbing with a minimum breaking strength of 5,000 pounds (22.2 kN). Connections between safety net panels shall be as strong as integral net components and be spaced no more than 6" (15 cm) apart.

Safety nets shall be installed with sufficient clearance underneath to prevent contact with the surface or structure below. When nets are used on bridges, the potential fall area from the walking/working surface to the net shall be unobstructed. Safety nets must extend outward from the outermost projection of the work surface as follows:

| SAFETY NETS *(cont.)* | |
|---|---|
| Vertical distance from working level to horizontal plane of net surface. | Minimum required horizontal distance of outer edge of net from edge of working surface. |
| Up to 5' (1.5 m) | 8' (2.4 m) |
| More than 5' (1.5 m) up to 10' (3 m) | 10' (3 m) |
| More than 10' (3 m) | 13' (3.9 m) |

Safety nets shall be capable of absorbing an impact force of a drop test consisting of a 400-pound (180 kg) bag of sand 30" (76 cm) in diameter dropped from the highest walking/working surface at which workers are exposed, but not from less than 42" (1.1 m) above that level. Items that have fallen into safety nets—including but not restricted to, materials, scrap, equipment and tools—must be removed as soon as possible and at least before the next work shift.

## WARNING LINES

Warning line systems consist of ropes, wires or chains, and supporting stanchions and are set up as follows:

- Flagged at not more than 6' (1.8-m) intervals with high-visibility material.

- Rigged and supported so that the lowest point including sag is no less than 34" (0.9 m) from the walking/working surface and its highest point is no more than 39" (1 m) from the walking/working surface.

- Stanchions, after being rigged with warning lines, shall be capable of resisting, without tipping over, a force of at least 16 pounds (71 N) applied horizontally against the stanchion, 30" (0.8 m) above the walking/working surface, perpendicular to the warning line and in the direction of the floor, roof or platform edge.

- Shall have a minimum tensile strength of 500 pounds (2.22 kN), and after being attached to the stanchions, must support without breaking the load applied to the stanchions as prescribed above.

- Shall be attached to each stanchion in such a way that pulling on one section of the line between stanchions will not result in slack being taken up in the adjacent section before the stanchion tips over.

Warning lines shall be erected around all sides of roof work areas. When mechanical equipment is being used, the warning line shall be erected not less than 6' (1.8 m) from the roof edge parallel to the direction of mechanical equipment operation, and not less than 10' (3 m) from the roof edge perpendicular to the direction of mechanical equipment operation.

When mechanical equipment is not being used, the warning line must be erected not less than 6' (1.8 m) from the roof edge.

# ROOFS AND WALL OPENINGS

## Low-Slope Roofs

Each employee engaged in roofing activities on low-slope roofs with unprotected sides and edges 6' (1.8 m) or more above lower levels shall be protected from falling by guardrail systems, safety net systems, personal fall arrest systems or a combination of a warning line system and guardrail system, warning line system and safety net system, warning line system and personal fall arrest system or warning line system and safety monitoring system. On roofs 50' (15.25 m) or less in width, the use of a safety monitoring system without a warning line system is permitted.

## Steep Roofs

Each employee on a steep roof with unprotected sides and edges 6' (1.8 m) or more above lower levels shall be protected by either guardrail systems with toeboards, a safety net system or a personal fall arrest system.

## Wall Openings

Each employee working on, at, above or near wall openings (including those with chutes attached) where the outside bottom edge of the wall opening is 6' (1.8 m) or more above lower levels and the inside bottom edge of the wall opening is less than 39" (1 m) above the walking/working surface must be protected from falling by the use of either a guardrail system, a safety net system or a personal fall arrest system.

## SIGNS

- Signs and symbols required by this subpart shall be visible at all times when work is being performed and shall be removed or covered promptly when the hazards no longer exist.

- Danger signs shall be used only where an immediate hazard exists. Danger signs shall have red as the predominating color for the upper panel, black outline on the borders and a white lower panel for additional sign wording.

- Caution signs shall be used only to warn against potential hazards or to caution against unsafe practices. Caution signs shall have yellow as the predominating color, black upper panel and borders, yellow lettering of "caution" on the black panel and the lower yellow panel for additional sign wording. Black lettering shall be used for additional wording.

Standard color of the background of signs shall be yellow, and the panel shall be black with yellow letters. Any letters used against the yellow background shall be black.

- Exit signs, when required, shall be lettered in legible red letters, not less than 6 inches (15.2 cm) high, on a white field and the principal stroke of the letters shall be at least 3/4 inch (1.9 cm) in width.

- Safety instruction signs. Safety instruction signs, when used, shall be white with green upper panel with white letters to convey the

## SIGNS *(cont.)*

principal message. Any additional wording on the sign shall be black letters on the white background.

- Directional signs, other than automotive traffic signs, shall be white with a black panel and a white directional symbol. Any additional wording on the sign shall be black letters on the white background.
- Construction areas shall be posted with legible traffic signs at points of hazard.

## WELDING OPERATIONS

The intense light associated with welding operations can cause serious and sometimes permanent eye damage if operators do not wear proper eye protection. The intensity of light or radiant energy produced by welding, cutting or brazing operations varies according to factors such as the task producing the light, the electrode size and the arc current. The following table shows the minimum protective shades for a variety of welding, cutting and brazing operations in general industry and in the shipbuilding industry.

## FILTER LENSES FOR PROTECTION AGAINST RADIANT ENERGY

| Operations | Electrode Size in $\frac{1}{32}$" (0.8 mm) | Arc Current | Minimum* Protective Shade |
|---|---|---|---|
| Shielded metal arc welding | <3 | <60 | 7 |
| | 3–5 | 60–160 | 8 |
| | 5–8 | 160–250 | 10 |
| | >8 | 250–550 | 11 |
| Gas metal arc welding and flux cored arc welding | | <60 | 7 |
| | | 60–160 | 10 |
| | | 160–250 | 10 |
| | | 250–500 | 10 |
| Gas tungsten arc welding | | <50 | 8 |
| | | 50–150 | 8 |
| | | 150–500 | 10 |
| Air carbon | (light) | <500 | 10 |
| Arc cutting | (heavy) | 500–1,000 | 11 |
| Plasma arc welding | | <20 | 6 |
| | | 20–100 | 8 |
| | | 100–400 | 10 |
| | | 400–800 | 11 |
| Plasma arc cutting | (light)** | <300 | 8 |
| | (medium)** | 300–400 | 9 |
| | (heavy)** | 400–800 | 10 |
| Torch brazing | | | 3 |
| Torch soldering | | | 2 |
| Carbon arc welding | | | 14 |

## FILTER LENSES FOR PROTECTION AGAINST RADIANT ENERGY (cont.)

| Operations | Plate Thickness (inches) | Plate Thickness (mm) | Minimum* Protective Shade |
|---|---|---|---|
| Gas welding: Light | <1/8 | <3.2 | 4 |
| Gas welding: Medium | 1/8–1/2 | 3.2–12.7 | 5 |
| Gas welding: Heavy | >1/2 | >12.7 | 6 |
| Oxygen cutting: Light | <1 | <25 | 3 |
| Oxygen cutting: Medium | 1–6 | 25–150 | 4 |
| Oxygen cutting: Heavy | >6 | >150 | 5 |

Source: 29 CFR 1910.133(a)(5).

\* As a rule of thumb, start with a shade that is too dark to see the weld zone. Then go to a lighter shade that gives sufficient view of the weld zone without going below the minimum. In oxyfuel gas welding or cutting where the torch produces a high yellow light, it is desirable to use a filter lens that absorbs the yellow or sodium line in the visible light of the (spectrum) operation.

\*\* These values apply where the actual arc is clearly seen. Experience has shown that lighter filters may be used when the arc is hidden by the workpiece.

The construction industry has separate requirements for filter lens protective levels for specific types of welding operations, as indicated in the table on the next page.

## CONSTRUCTION INDUSTRY REQUIREMENTS FOR FILTER LENS SHADE NUMBERS FOR PROTECTION AGAINST RADIANT ENERGY

| Welding Operation | Shade Number |
|---|---|
| Shielded metal-arc welding<br>$\frac{1}{16}$", $\frac{3}{32}$", $\frac{1}{8}$", $\frac{5}{32}$" diameter electrodes | 10 |
| Gas-shielded arc welding (non-ferrous)<br>$\frac{1}{16}$", $\frac{3}{32}$", $\frac{1}{8}$", $\frac{5}{32}$" diameter electrodes | 11 |
| Gas-shielded arc welding (ferrous)<br>$\frac{1}{16}$", $\frac{3}{32}$", $\frac{1}{8}$", $\frac{5}{32}$" diameter electrodes | 12 |
| Shielded metal-arc welding<br>$\frac{3}{16}$", $\frac{7}{32}$", $\frac{1}{4}$" diameter electrodes | 12 |
| $\frac{5}{16}$", $\frac{3}{8}$" diameter electrodes | 14 |
| Atomic hydrogen welding | 10–14 |
| Carbon-arc welding | 14 |
| Soldering | 2 |
| Torch brazing | 3 or 4 |
| Light cutting, up to 1" | 3 or 4 |
| Medium cutting, 1" to 6" | 4 or 5 |
| Heavy cutting, more than 6" | 5 or 6 |
| Gas welding (light), up to $\frac{1}{8}$" | 4 or 5 |
| Gas welding (medium), $\frac{1}{8}$" to $\frac{1}{2}$" | 5 or 6 |
| Gas welding (heavy), more than $\frac{1}{2}$" | 6 or 8 |

## LASER OPERATIONS

Laser light radiation can be extremely dangerous to the unprotected eye, and direct or reflected beams can cause permanent eye damage. Laser retinal burns can be painless, so it is essential that all personnel in or around laser operations wear appropriate eye protection.

Laser safety goggles should protect for the specific wavelength of the laser and must be of sufficient optical density for the energy involved. Safety goggles intended for use with laser beams must be labeled with the laser wavelengths for which they are intended to be used, the optical density of those wavelengths and the visible light transmission.

The table below lists maximum power or energy densities and appropriate protection levels for optical densities 5 through 8.

## SELECTING LASER SAFETY GLASS

| Intensity, CW Maximum Power Density (watts/cm$^2$) | Attenuation | |
|---|---|---|
| | Optical Density | Attenuation Factor |
| 10-2 | 5 | $10^5$ |
| 10-1 | 6 | $10^6$ |
| 1.0 | 7 | $10^7$ |
| 10.0 | 8 | $10^8$ |

### (b) Branch circuits-(1) Ground-fault protection

**(i) General.** The employer shall use either ground-fault circuit interrupters as specified in paragraph (b)(1)(ii) of this section or an assured equipment grounding conductor program as specified in paragraph (b)(1)(iii) of this section to protect employees on construction sites. These requirements are in addition to any other requirements for equipment grounding conductors.

**(ii) Ground-fault circuit interrupters.** All 120-volt, single-phase, 15- and 20-ampere receptacle outlets on construction sites, which are not a part of the permanent wiring of the building or structure and are in use by employees, shall have approved ground-fault circuit interrupters for personnel protection. Receptacles on a two-wire, single-phase portable or vehicle-mounted generator rated not more than 5kW, where the circuit conductors of the generator are insulated from the generator frame and all other grounded surfaces, need not be protected with ground-fault circuit interrupters.

**(iii) Assured equipment-grounding conductor program.** The employer shall establish and implement an assured equipment-grounding conductor program on construction sites covering all cord sets, receptacles that are not a part of the building or structure and equipment connected by cord and plug that are available for use or used by employees. This program shall comply with the following minimum requirements:

(A) A written description of the program, including the specific procedures adopted by the employer, shall be available at the job site for inspection and copying by the assistant secretary and any affected employee.

(B) The employer shall designate one or more competent persons to implement the program.

(C) Each cord set, attachment cap, plug and receptacle of cord sets, and any equipment connected by cord and plug, except cord sets and receptacles that are fixed and not exposed to damage, shall be visually inspected before each day's use for external defects, such as deformed or missing pins or insulation damage, and for indications of possible internal damage. Equipment found to be damaged or defective shall not be used until repaired.

(D) The following tests shall be performed on all cord sets, receptacles that are not a part of the permanent wiring of the building or structure and cord- and plug-connected equipment required to be grounded:

(1) All equipment-grounding conductors shall be tested for continuity and shall be electrically continuous.

(2) Each receptacle and attachment cap or plug shall be tested for correct attachment of the equipment grounding conductor. The equipment-grounding conductor shall be connected to its proper terminal.

(E) All required tests shall be performed:

(1) Before first use;

(2) Before equipment is returned to service following any repairs;

(3) Before equipment is used after any incident that can be reasonably suspected to have caused damage (e.g., when a cord set is run over); and

(4) At intervals not to exceed 3 months, except that cord sets and receptacles that are fixed and not exposed to damage shall be tested at intervals not exceeding 6 months.

(F) The employer shall not make available or permit the use by employees of any equipment that has not met the requirements of paragraph (b)(1)(iii) of this section.

(G) Tests performed as required in this paragraph shall be recorded. This test record shall identify each receptacle, cord set, and cord- and plug-connected equipment that passed the test and shall indicate the last date on which it was tested or the interval for which it was tested. This record shall be kept by means of logs, color coding or other effective means and shall be maintained until replaced by a more current record. The record shall be made available on the job site for inspection by the assistant secretary and any affected employee.

# CHAPTER 9
## *Company Safety Plan*

This chapter gives a detailed review of each of a company's overall safety plan which includes safety policies and goals, the responsibilities of all employees in a supervisory capacity, fire prevention, hazardous conditions and materials as well as educational and training programs for all employees.

### SAMPLE POLICY STATEMENT

[COMPANY NAME]
will preserve the safety and health of all employees. We will provide the resources necessary to manage, control or eliminate safety and health hazards. We will not ignore on-the-job threats to the safety or health of our employees.

All employees are responsible for working safely and productively, as well as to recognize and inform the company of hazards in their work areas.

Employees are also responsible for following safe work practices, including the use of personal protective equipment where necessary.

_____

[Company Name] President

## GOAL

The primary goal of [COMPANY NAME] is to operate a profitable business by serving its customers.

A primary element of reaching this goal is to keep our employees free from injuries, illness or harm on the job. We will achieve this, in part, by delegating responsibility and accountability to all involved in this company's operation.

Our safety goals are the following:

- Minimize or eliminate all injuries and accidents.
- Minimize loss of property and equipment.
- Eliminate all OSHA fines.
- Reduce workers' compensation costs.
- Reduce operating costs.

## SPECIFIC STEPS

To achieve our safety goals, we will do the following:

- Appoint well trained people to be our safety coordinators.
- Provide all necessary safety training, especially to safety coordinators.
- Establish company safety goals.
- Secure honest safety feedback and information from our job sites. All employees must be able to keep us informed as to safety and health threats.
- Adapt company actions as required to meet safety objectives.

## SPECIFIC STEPS *(cont.)*

- Develop and implement a written safety and health program.

- Hold all employees accountable for performance of safety responsibilities.

- Reviewing the safety and health program annually. Revising or update as required.

## SAFETY MANAGER

A safety manager shall be appointed to review all safety issues with both field and office personnel.

- Gather relevant safety information from all sources.

- Discuss safety policies and procedures with all involved.

- Make recommendations for improvements.

- Review accident investigation reports on all accidents and near misses.

- Identify unsafe conditions and work practices and enforce corrections.

The safety manager is authorized to shut down projects, without consultation, upon discovery of any serious threat to our employees.

## NOTICE TO ALL EMPLOYEES

_____ has been designated as our safety manager. [His/her] cell phone and office phone numbers are:

Office: (000) 000-0000

Cell: (000) 000-0000

It is the duty of the safety coordinator to assist all of you in keeping our jobs safe.

Please contact _____ immediately regarding any on-the-job health or safety issues. This is one of _____'s primary areas of responsibility.

Our safety managers are also responsible for:

- Introducing our safety program to new employees.
- Following up on suggestions made by employees and documenting suggestions and responses.
- Assisting personnel in the execution of safety policies.
- Conducting safety inspections periodically.
- Addressing all hazards or potential hazards as needed.
- Preparing monthly accident reports and investigations.
- Maintaining an adequate stock of first aid supplies and other safety equipment to ensure availability.
- Making sure there is an adequate number of employees who are certified in first aid.
- Staying current with OSHA regulations and local safety mandates.

## SUPERVISOR/FOREMAN

It is the responsibility of our supervisors and foremen to establish an operating atmosphere that ensures that safety and health is managed carefully.

Supervisors and foremen are required to do the following:

- Regularly emphasize that accident and health hazard exposure prevention are a condition of employment.
- Identify operational oversights that could contribute to accidents.
- Participate in safety- and health-related activities, including attending safety meetings, attending reviews of the facility and correcting employee behavior that can result in accidents and injuries.
- Spend time with each person hired to explain the safety policies and the hazards of his/her particular work.
- Making sure that a competent person is present as required.
- Prevent any sacrifice of safety for expediency, nor allow workers to do so.
- Enforce safety rules consistently. Follow the company's discipline and enforcement procedures.
- Conduct daily job-site safety inspections and correct safety violations.

## EMPLOYEE RESPONSIBILITIES

It is the duty of each and every employee to know the safety rules and to conduct his/her work in compliance with these rules. Disregard of the safety and health rules shall be grounds for disciplinary action up to and including termination. It is also the duty of all employees to make full use of the safeguards provided for their protection. Every employee will receive safety instructions and a copy of the company safety and health program upon hire.

Employee must do the following:

- Read, understand and follow safety and health rules and procedures.
- Wear personal protective equipment at all times when working in areas where there is a danger of injury.
- Wear suitable work clothes as determined by the supervisor or foreman.
- Perform all tasks safely, as directed by a supervisor or foreman.
- Report all injuries to a supervisor or foreman, and seek treatment promptly.
- Know the location of first aid, fire fighting equipment and other safety devices.
- Attend all required safety and health meetings.
- Do not perform potentially hazardous tasks, or use any hazardous material unless properly trained to do so. Follow all safety procedures.
- Stop and ask questions if in doubt about the safety of any operation.

## DISCIPLINE AND ENFORCEMENT

[COMPANY NAME] maintains standards of employee conduct and supervisory practices that support and promote effective and safe business operations. These supervisory practices include administering corrective action when employee safety performance or conduct jeopardizes this goal. This policy sets forth general guidelines for a corrective action process aimed to document and correct undesirable employee behavior. Major elements of this policy include:

A. Constructive criticism/instruction by the employee's supervisor/foreman to educate and inform the employee of appropriate safety performance and behavior.

B. Correcting the employee's negative behavior to the extent required.

C. Informing the employee that continued violation of company safety policies can result in termination.

D. Written documentation of disciplinary warnings and corrective action taken.

Depending on the facts and circumstances involved with each situation, the company may choose any corrective action including immediate termination. However, in most circumstances the following steps will be followed:

1. **Verbal Warning** informally documented by supervisor, foreman or safety manager, for minor

## DISCIPLINE AND ENFORCEMENT *(cont.)*

infractions of company safety rules. Supervisor, foreman or safety manager must inform the employee what safety rule or policy was violated and how to correct the problem.

2. **Written Warning** documented in the employee's file. Repeated minor infractions or a more substantial safety infraction requires issuance of a written warning. The employee should acknowledge the warning by signing the document before it is placed in their personnel file.

3. **Suspension** for three (3) working days if the employee fails to appropriately respond or if management determines the infraction was sufficient for a suspension.

4. **Termination** for repeated or serious safety infractions.

## CONTROL OF HAZARDS

Where feasible, workplace hazards are prevented by effective design of the job site or job. Where it is not feasible to eliminate such hazards, they must be controlled to prevent unsafe and unhealthy exposure. Once a potential hazard is recognized, the elimination or control must be done in a timely manner. These procedures include measures such as the following:

- Maintaining all extension cords and equipment.
- Ensuring all guards and safety devices are working.
- Periodically inspecting the worksite for safety hazards.

## CONTROL OF HAZARDS *(cont.)*

- Establishing a medical program that provides applicable first aid to the site, as well as nearby physician and emergency phone numbers.
- Addressing any and all safety hazards with employees.

## FIRE PREVENTION

Fire prevention is an important part of protecting employees and company assets. Fire hazards must be controlled to prevent unsafe conditions. Once a potential hazard is recognized, it must be eliminated or controlled in a timely manner. The following fire prevention requirements must be met for each site:

- One conspicuously located 2A fire extinguisher (or equivalent) for every floor.
- One 2A conspicuously located fire extinguisher (or equivalent) for every 3,000 sq/ft.
- A conspicuously located, 10B fire extinguisher for everywhere more than 5 gallons of flammable liquids or gas are stored.
- Generators and internal combustion engines located away from combustible materials.
- Site free from accumulation of combustible materials or weeds.
- No obstructions or combustible materials piled in the exits.
- No more than 25 gallons of combustible liquids stored on site.
- No *LPG* containers stored in any buildings or enclosed spaces.
- Fire extinguishers in the immediate vicinity where welding, cutting or heating is being done.

## TRAINING AND EDUCATION

Training is an essential component of an effective safety and health program that addresses the responsibilities of both management and employees at the site. Training is most effective when incorporated into other education on performance requirements and job practices. Training programs should be provided as follows:

- Initially when a safety and health plan is developed.
- For all new employees before beginning work.
- When new equipment, materials or processes are introduced.
- When procedures have been updated or revised.
- When experiences/operations show that employee performance must be improved.
- At least annually.

Besides the standard training, employees should also be trained in the recognition of hazards, or the ability to look at an operation and identify unsafe acts and conditions. A list of typical hazards employees should be able to recognize may include:

- **Fall Hazards:** falls from floors, roofs and roof openings, ladders (straight and step), scaffolds, wall openings, tripping, trenches, steel erection, stairs, chairs.

## TRAINING AND EDUCATION *(cont.)*

- **Electrical Hazards:** appliances, damaged cords, outlets, overloads, overhead high voltage, extension cords, portable tools (broken casing or damaged wiring), grounding, metal boxes, switches, ground fault interrupters.
- **Housekeeping Issues:** exits, walkways, floors, trash, storage of materials (hazardous and non-hazardous), protruding nails, etc.
- **Fire Hazards:** oily-dirty rags, combustibles, fuel gas cylinders, exits (blocked), trips/slips stairs, uneven flooring, electrical cords, icy walkways.
- **Health Hazards:** silicosis, asbestos, loss of hearing, eye injury due to flying objects.

Employees trained in the recognition and reporting of hazards and supervisors/foremen trained in the correction of hazards will substantially reduce the likelihood of a serious injury.

## EMERGENCY RESPONSE TO HAZARDOUS SUBSTANCES

If any substance is found of unknown origin, **LEAVE IT ALONE!** Immediately evacuate the area and contact the nearest hazardous material response team. Do not allow employees on site until declared safe by the response team.

### First Aid
- Arrangements must be made BEFORE starting the project to provide for prompt medical response in the event of an emergency.

## EMERGENCY RESPONSE TO
## HAZARDOUS SUBSTANCES *(cont.)*

- In areas where severe bleeding, suffocation or electrical shock can occur, a 3 to 4-minute response time is required.
- If medical attention is not available within 4 minutes, then a person trained in first aid must be available on the job site at all times.
- An appropriate, weatherproof first aid kit must be on-site, and must be checked weekly.
- Provisions for an ambulance or other transportation must be made in advance.
- Contact methods must be provided.
- Telephone numbers must be posted where 911 is not available.

The following person has adequate training to render first aid in the event of a medical emergency in areas where emergency response time is in excess of 4 minutes:_____

First aid kits are located at the following locations:
- _____
- _____

Every employee shall be trained in the following emergency procedures:

- Evacuation plan
- Alarm systems
- Shutdown procedures for equipment
- Types of potential emergencies

It is the employer's responsibility to review its job sites and address all potential emergency situations.

## EMERGENCY RESPONSE PLAN

(Must be filled out BEFORE beginning work on each site.)

**Work Site**

Job: _____

Street Name: _____

Address: _____

Address: _____

Job Phone Contact: _____

**Emergency Numbers**

Fire Dept/EMS: _____

Ambulance: _____

Hospital/Clinic: _____

_____

_____

## RECORDKEEPING AND OSHA LOG REVIEW

In the event of a fatality (death on the job) or catastrophe (accident resulting in hospitalization of three or more workers) contact [safety manager] at the office or via cell-phone:

Office: [(000) 000-0000]
Cell: [(000) 000-0000]

The safety manager will in turn report it to the OSHA Regional Office within 8 hours after the occurrence.

If an injury or accident should ever occur, you are to report it to your supervisor or foreman as soon as possible. A log entry and summary report shall be maintained for every recordable injury and illness. The entry should be done within 7 days after the injury or illness has occurred. An appropriate form shall be used for the recording. A recordable injury or illness is any injury resulting in loss of consciousness, days away from work, days of restricted work or medical treatment beyond first aid.

First aid includes:

- Tetanus shots.
- Band-aids or butterfly bandages.
- Cleaning, flushing or soaking wounds.
- Ace bandages and wraps.
- Non-prescription drugs at non-prescription strength (aspirin, acetaminophen, etc.).
- Drilling fingernails/toenails.
- Eye patches, eye flushing and foreign body removal from eye with a cotton swab.
- Finger guards.
- Hot or cold packs.
- Drinking fluids for heat stress.

An annual summary of recordable injuries and illnesses must be posted at a conspicuous location in the workplace and contain the following information: calendar year, company name, establishment name, establishment address, certifying signature, title and date. If no injury or illness occurred in the year, zeroes must be entered on the total line.

The OSHA logs should be evaluated by the employer to determine trends or patterns in injuries in order to appropriately address hazards and implement prevention strategies.

## ACCIDENT INVESTIGATION

### Supervisors/Foreman
- Provide first aid and call for emergency medical care if required.
- If further medical treatment is required, arrange to have an employer representative accompany the injured employee to the medical facility.
- Secure area, equipment and personnel from injury and further damage.
- Contact safety manager.

### Safety Manager
- Investigate the incident (injury). Gather facts and employee and witness statements; take pictures and physical measurements of incident site and equipment involved.
- Complete an incident investigation report form and the necessary workers' compensation paperwork within 24 hours whenever possible.
- Ensure that corrective action to prevent a recurrence is taken.
- Discuss the incident, if appropriate, in safety and other employee meetings with the intent to prevent a recurrence.
- Discuss incident with other supervisors, foremen and other management personnel.

## ACCIDENT INVESTIGATION *(cont.)*

- If the injury warrants time away from work, be sure that the absence is authorized by a physician and that you maintain contact with your employee while he/she remains off work.
- Monitor status of employee(s) off work, maintain contact with employee and encourage return to work even if restrictions are imposed by his/her physician.
- When injured employee(s) return to work they should not be allowed to return to work without "return to work" release forms from the physician. Review the release carefully and ensure that you can accommodate the restrictions, and that the employee follows the restrictions indicated by the physician.

## SAFETY RULES AND PROCEDURES

- No employee is expected to undertake a job until that person has received adequate training.
- All employees shall be trained on every potential hazard that they could be exposed to and how to protect themselves.
- Employees are not required to work under conditions that are unsanitary, dangerous or hazardous to their health.
- Only qualified trained personnel are permitted to operate machinery or equipment.
- All injuries must be reported to your supervision/foreman.
- Manufacturer's specifications/limitations/instructions shall be followed.

## SAFETY RULES AND PROCEDURES *(cont.)*

- Particular attention should be given to new employees and to employees moving to new jobs or performing non-routine tasks.
- Emergency numbers shall be posted and reviewed with employees.
- Each employee working in an excavation or trench shall be protected from cave-ins by an adequate protective system.
- Employees working in areas where there is a danger of head injury, excessive noise exposure or potential eye and face injury shall be protected by personal protection equipment.
- All hand and power tools and similar equipment, whether furnished by the employer or the employee, shall be maintained in a safe condition.
- All materials stored in tiers shall be stacked, racked, blocked, interlocked or otherwise secured to prevent sliding, falling or collapse.
- The employer shall ensure that electrical equipment is free from recognized hazards that are likely to cause death or serious physical harm to employees.
- All scaffolding shall be erected in accordance with the CFR 1926.451 subpart L. standard. Guardrails for fall protection and ladders for safe access shall be used.
- Places of employment shall be kept clean and the floor of every workroom shall be maintained, so far as practicable, in a dry condition; standing water shall be removed. Where wet processes are used, drainage shall be

maintained and false floors, platforms, mats or other dry standing places or appropriate water-proof footgear shall be provided.

- To facilitate cleaning, every floor, working place and passageway shall be kept free from protruding nails, splinters, loose boards and holes and openings.
- All floor openings, open sided floor and wall openings shall be guarded by standard railings and toe boards or cover.
- The employer shall comply with the manufacturer's specifications and limitations applicable to the operation of any and all cranes and derricks.
- All equipment left unattended at night, adjacent to a highway in normal use or adjacent to construction areas where work is in progress, shall have appropriate lights or reflectors, or barricades equipped with appropriate lights or reflectors, to identify the location of the equipment.
- No construction loads shall be placed on a concrete structure or portion of a concrete structure unless the employer determines, based on information received from a person who is qualified in structural design, that the structure or portion of the structure is capable of supporting the loads.
- A stairway or ladder shall be provided at all personnel points of access where there is a break in elevation of 19" or more, and no ramp, runway, sloped embankment or personnel hoist is provided.

## EMPLOYEE EMERGENCY ACTION PLAN — FIRE AND OTHER EMERGENCIES

The following emergency action plan is appropriate only for small construction sites. Larger sites should have a much more detailed plan.

1. **Emergency escape procedures:** Immediately leave the building through the closest practical exit. Meet up at the foremen's truck.

2. **Critical plant operations:** Shut off the generator on your way out if possible, otherwise evacuate the building.

3. **Accounting for employees:** Foreman/Supervisor is to account for all employees after emergency evacuation has been completed and assign duties as necessary.

4. **Means of reporting fires and other emergencies:** Dial 911 on the cell phone, report the location of the emergency and provide directions to the responders.

5. **Further information:** Contact the safety coordinator of further information or explanation of duties under the plan.

## EMPLOYEE EMERGENCY ACTION PLAN — FIRE AND OTHER EMERGENCIES *(cont.)*

**Training:** Before implementing the emergency action plan, a sufficient number of persons to assist in the safe and orderly emergency evacuation of employees will be designated and trained. The plan will be reviewed with each employee covered by the plan at the following times:

1. Initially when the plan is developed or upon initial assignment.

2. Whenever the employee's responsibilities or designated actions under the plan change.

3. Whenever the plan is changed.

The plan will be kept at the work site and made available for employee review. For those employers with 10 or fewer employees, the emergency action plan may be communicated orally to employees and the employer need not maintain a written plan.

## INJURY AND ILLNESS INCIDENT REPORT

Completed by _____

Title _____

Phone (____)____-_____ Date ____/____/____

### Information about the employee

1) Full name _____

2) Street _____

   City_____ State_____ ZIP_____

3) Date of birth _____/_____/_____

4) Date hired _____/_____/_____

5) ❏ Male

   ❏ Female

### Information about the physician or other health care professional

6) Name of physician or other health care professional _____

_____

7) If treatment was given away from the worksite, where was it given?

   Facility _____

   Street _____

   City _____ State _____ ZIP_____

8) Was employee treated in an emergency room?

   ❏ Yes

   ❏ No

## INJURY AND ILLNESS INCIDENT REPORT *(cont.)*

9) Was employee hospitalized overnight as an in-patient?

❒ Yes

❒ No

### Information about the case

10) Case number from the *Log* _____ *(Transfer the case number from the Log after you record the case.)*

11) Date of injury or illness _____/_____/_____

12) Time employee began work _____ AM/PM

13) Time of event _____ AM/PM

   ❒ Check if time cannot be determined

14) **What was the employee doing just before the incident occurred**? Describe the activity, as well as the tools, equipment or material the employee was using. Be specific. *Examples*: "climbing a ladder while carrying roofing materials"; "spraying chlorine from hand sprayer"; "daily computer key-entry."

_____

_____

## INJURY AND ILLNESS INCIDENT REPORT *(cont.)*

15) **What happened?** Tell us how the injury occurred. *Examples*: "When ladder slipped on wet floor, worker fell 20 feet"; "Worker was sprayed with chlorine when gasket broke during replacement"; "Worker developed soreness in wrist over time."

_____

_____

16) **What was the injury or illness?** Tell us the part of the body that was affected and how it was affected; be more specific than "hurt," "pain" or sore." *Examples*: "strained back"; "chemical burn, hand"; "carpal tunnel syndrome."

_____

_____

17) **What object or substance directly harmed the employee?** *Examples*: "concrete floor"; "chlorine"; "radial arm saw." *If this question does not apply to the incident, leave it blank.*

_____

_____

18) **If the employee died, when did death occur?**

Date of death _____/_____/_____

## SUMMARY OF WORK-RELATED
## INJURIES AND ILLNESSES

**Number of Cases**

| Total number of deaths | Total number of cases with days away from work | Total number of cases with job transfer or restriction | Total number of other recordable cases |
|---|---|---|---|
| _____ | _____ | _____ | _____ |

**Number of Days**

| Total number of days of job transfer or restriction | Total number of days away from work |
|---|---|
| _____ | _____ |

**Injury and Illness Types**
(M) Total number of . . .
(1) Injuries                                    _____
(2) Skin disorders                          _____
(3) Respiratory conditions             _____
(4) Poisonings                               _____
(5) All other illnesses                    _____

*Post this Summary page from February 1 to April 30 of the year following the year covered by the form.*

## SUMMARY OF WORK-RELATED INJURIES AND ILLNESSES *(cont.)*

**Company Information**

Company name _____

Street _____

City _____ State _____ ZIP _____

Phone _____

Web-site _____

e-mail _____

Industry description *(e.g., Manufacture of motor truck trailers)*

_____

Standard Industrial Classification (SIC), if known *(e.g., SIC 3715)*

_____

Annual average number of employees _____

Total hours worked by all employees last year _____

## SUMMARY OF WORK-RELATED INJURIES AND ILLNESSES *(cont.)*

I have read and understand the attached company policies and procedures and agree to abide by them. I have also had the duties of the position that I have accepted explained to me, and I understand the requirements of the position. I understand that any violation of the above policies is reason for disciplinary action up to and including termination.

_____     _____
Employee Signature                                Date

I certify that I have examined this document and that to the best of my knowledge the entries are true, accurate and complete.

Company executive _____

Title _____

Phone (____)____-____ Date ____/____/____

_____     _____
Executive Signature                               Date

**Knowingly falsifying this document may result in a fine.**

# CHAPTER 10
## *Glossary and Abbreviations*

## GLOSSARY

### A

**ABS:** Abbreviation for acrylonitrile butadiene styrene; a plastic pipe used for plumbing construction.

**Abut:** To join end-to-end.

**Accelerator:** A concrete additive used to speed the curing time of freshly poured concrete.

**Acoustical:** Referring to the study of sound transmission or reduction.

**Adhesive:** A bonding material used to bond two materials together.

**Adjacent:** Touching; next to.

**Aggregate:** Fine, lightweight, coarse or heavyweight grades of sand, vermiculite, perlite or gravel added to cement for concrete or plaster.

**Air-Drying:** A method of removing excess moisture from lumber using natural circulation of air.

**Air Handling Unit:** A mechanical unit used for air conditioning or movement of air as in direct supply or exhaust of air within a structure.

**Allowable Load:** Maximum supportable load of any construction component(s).

**Allowable Span:** The maximum length permissible for any framing component without support.

**Anchorage:** A secure point of attachment for lifelines, lanyards or deceleration devices.

**Anchor Bolt:** A J- or L-shaped steel rod threaded on one end for securing structural members to concrete or masonry.

**Anchored Bridging:** Steel joist bridging connected to a bridging terminus point.

**Apron:** A piece of window trim that is located beneath the window sill; also used to designate the front of a building, such as the concrete apron in front of a garage.

**Arbor:** An axle on which a cutting tool is mounted; it is a common term used in reference to the mounting of a circular saw blade.

**Architect's Scale:** A rule with scales indicating feet, inches and fractions of inches.

**Asphalt:** The general term for a black material produced as a by-product of oil (asphalt) or coal (pitch or coal tar).

**Asphalt Shingle:** A composition-type shingle used on a roof and made of a saturated felt paper with ground-up pieces of stone embedded and held in place by asphaltum.

**Asphalt Shingles:** Shingles made of asphalt or tar-impregnated paper with a mineral material embedded; fire resistant.

**Awl:** A tool used to mark wood with a scratch mark; can be used to produce pilot holes for screws.

**Awning Window:** A window that is hinged at the top and the bottom swings outward.

## B

**Backfill:** Any deleterious material (sand, gravel, etc.) used to fill an excavation.

**Backhoe:** Self-powered excavation equipment that digs by pulling a boom mounted bucket toward itself.

**Backsplash:** The vertical part of a countertop that runs along the wall to prevent splashes from marring the wall.

**Balloon Framing:** Wall construction extending from the foundation to the roof structure without interruption; used in residential construction only.

**Baluster:** The part of the staircase that supports the handrail or bannister.

**Balustrade:** A complete handrail assembly; includes the rails, balusters, subrails and fillets.

**Bank Plug:** Piece of lumber (usually 2" × 4") driven into the ground to stand some distance, usually 24", above ground level; surveyors place nails in the bank plugs a given distance above the road surface so a string line can be stretched between the plugs to measure grade.

**Bannister:** The part of the staircase that fits on top of the balusters.

**Baseboard:** Molding covering the joint between a finished wall and the floor.

**Base Shoe:** A molding added at the bottom of a baseboard; used to cover the edge of finished flooring or carpeting.

**Batt Insulation:** An insulating material formed into sheets or rolls with a foil or paper backing; to be installed between framing members.

**Batten:** A narrow piece of wood used to cover a joint.

**Batter Boards:** Boards used to frame in the corners of a proposed building during layout and excavating work.

**Beam:** A horizontal framing member; it may be made of steel or wood and usually refers to a wooden beam at least 5" thick and at least 2" wider than it is thick.

**Bearing Partition:** An interior divider or wall that supports the weight of the structure above it.

**Bearing Wall:** A wall having weight-bearing properties associated with holding up a building's roof or second floor.

**Benching:** Making steplike cuts into a slope; used for erosion control or to tie a new fill into an existing slope.

**Benchmark:** The point of known elevation from which the surveyors can establish all their grades.

**Berm:** A raised earth embankment; the shoulder of a paved road; the area between the curb and the gutter and a sidewalk.

**Bevel:** A tool that can be adjusted to any angle; it helps make cuts at the number of degrees that is desired and is a good device for transferring angles from one place to another.

**Bibb:** Also known as a *hose bibb*; a faucet used to connect a hose.

**Bi-fold:** A double-leaf door used primarily for closet doors in residential construction.

**Bird Mouth:** A notch cut into a roof rafter so that it can rest smoothly on the top plate.

**Bitumen:** The general term used to identify asphalt and coal tar.

**Blistering:** The condition that paint presents when air or moisture is trapped underneath and makes bubbles that break into flaky particles and ragged edges.

**Blocking:** A piece of wood fastened between structural members to strengthen them; generally solid or cross-tie wood or metal cross-tie members to perform the same task.

**Board Foot:** A unit of lumber measure equaling 144 cubic inches; the base unit is 1" thick and 12" square or 1 × 12 × 12 = 144 cubic inches.

**Body Belt:** A strap with means both for securing it

about the waist and for attaching it to a lanyard, lifeline or deceleration device.

**Body Harness:** Straps that may be secured about the person in a manner that distributes the fall-arrest forces over at least the thighs, pelvis, waist, chest and shoulders with a means for attaching the harness to other components of a personal fall arrest system.

**Bolted Diagonal Bridging:** Diagonal bridging that is bolted to a steel joist or joists.

**Bond:** In masonry, the interlocking system of brick or block to be installed.

**Borrow Site:** An area from which earth is taken for hauling to a job site that is short of earth needed to build an embankment.

**Bottom Or Heel Cut:** The cutout of the rafter end that rests against the plate; also called the *foot* or *seat cut*.

**Bow:** A term used to indicate an upward warp along the length of a piece of lumber that is laid.

**Bow Window:** A window unit that projects from an exterior wall.

**Brace:** An inclined piece of lumber applied to a wall or to roof rafters to add strength.

**Bridging Clip:** A device that is attached to the steel joist to allow the bolting of the bridging to the steel joist.

**Bridging Terminus:** A wall, a beam, tandem joists (with all bridging installed and a horizontal truss in the plane of the top chord) or other element at an end or intermediate point(s) of a line of bridging that provides an anchor point for the steel joist bridging.

**Bridging:** Used to keep joists from twisting or bending.

**Builder's Level:** A tripod-mounted device that uses optical sighting to make sure that a straight line is sighted and that the reference point is level.

**Building Codes:** Rules and regulations that are formulated in a code by a local housing authority or governing body.

**Building Paper:** Also called tar paper, roofing paper and a number of other terms; paper having a black coating of asphalt for use in weatherproofing.

**Building Permits:** A series of permits that must be obtained for building; allows for inspections of the work and for placing the house on the tax roles.

**Bull Float:** A tool used to spread out and smooth concrete.

**Butt:** To meet edge to edge, such as in a joining of wooden edges.

## C

**Calcium Chloride:** A concrete admixture used for accelerating the cure time.

**California Bearing Ratio:** A system used for determining the bearing capacity of a foundation.

**Carriage:** A notched stair frame.

**Casement:** A type of window hinged to swing outward.

**Casing:** The trim that goes on around the edge of a door opening; also the trim around a window.

**Catch Basin:** A complete drain box made in various depths and sizes; water drains into a pit, then through a pipe connected to the box.

**Catch Point:** Another name for hinge point or top of shoulder.

**Caulk:** Any material used to seal walls, windows and doors to keep out the weather.

**Cement:** A material that, when combined with water, hardens due to chemical reaction; the basis for a concrete mix.

**Cement Plaster:** A mixture of gypsum, cement, hydrated lime, sand and water; used primarily for exterior wall finish.

**Cementitious:** Able to harden like cement.

**Center Line:** The point on stakes or drawings that indicates the halfway point between two sides.

**Chair:** Small device used to support horizontal rebar prior to the concrete placement.

**Choker:** A wire rope or synthetic fiber rigging assembly used to attach a load to a hoisting device.

**Chord:** Top or bottom member of a truss.

**Clear And Grub:** To remove all vegetation, trees, concrete or anything that will interfere with construction inside the limits of the project.

**Cleat:** A ladder crosspiece of rectangular cross section placed on edge on which a person may step while ascending or descending a ladder.

**Cleat:** Any strip of material attached to the surface of another material to strengthen, support or secure a third material.

**Cold Forming:** The process of using press brakes, rolls or other methods to shape steel into desired cross sections at room temperature.

**Collar Tie:** Horizontal framing member tying the raftering together above the plate line.

**Column:** A load-carrying vertical member that is part of the primary skeletal framing system; columns do not include posts.

**Common Rafter:** A structural member that extends without interruption from the ridge to the plate line in a sloped roof structure.

**Compactor:** A machine for compacting soil; can be pulled or self powered.

**Competent Person:** One who is capable of identifying existing and predictable hazards in the surroundings or working conditions that are unsanitary, hazardous or dangerous to employees, and who has authorization to take prompt corrective measures to eliminate them.

**Computer-Aided Drafting:** Computer-aided design and drafting.

**Concrete:** A mixture of sand, gravel and cement in water.

**Condensation:** The process by which moisture in the air becomes water or ice on a surface (such as a window) whose temperature is colder than the air's temperature.

**Connector:** An employee who, working with hoisting equipment, is placing and connecting structural members and/or components.

**Connector:** A device that is used to couple (connect) parts of a personal fall arrest system or positioning device system together.

**Constructibility:** The ability to erect structural steel members in accordance with subpart R without having to alter the over-all structural design.

**Construction Load:** Any load other than the weight of the employee(s), the joists and the bridging bundle in joist erection.

**Contour Line:** Solid or dashed lines showing the elevation of the earth on a project.

**Controlled Access Zone:** A work area designated and clearly marked in which certain types of work (such as overhand bricklaying) may take place without the use of conventional fall protection systems—guardrail, personal arrest or safety net—to protect the employees working in the zone.

**Controlled Decking Zone:** An area in which certain work (for example, initial installation and placement of metal decking) may take place without the use of guardrail systems, personal fall arrest systems, fall restraint systems or safety net systems and where access to the zone is controlled.

**Controlled Load Lowering:** Taking down a load by means of a mechanical hoist drum device that allows a hoisted load to be lowered with maximum control using the gear train or hydraulic components of the hoist mechanism; controlled load lowering requires the use of the hoist drive motor, rather than the load hoist brake, to lower the load.

**Controlling Contractor:** A prime contractor, general contractor, construction manager or any other legal entity who has the overall responsibility for the construction of the project— its planning, quality and completion.

**Convection:** The transfer of heat through the movement of a liquid or gas.

**Corner Beads:** Metal strips that prevent damage to drywall corners.

**Cornice:** The area under the roof overhang; usually enclosed or boxed in.

**Crawl Space:** The area under a floor that is not fully excavated; only excavated sufficiently to allow one to crawl under it to get at the electrical or plumbing devices.

**Cripple Jack:** A jack rafter with a cut that fits in between a hip and a valley rafter.

**Cripple Rafter:** Not as long as the regular rafter; used to span a given area.

**Cripple Stud:** A short stud that fills out the position where the stud would have been located if a window, door or some other opening had not been there.

**Critical Lift:** A lift that (1) exceeds 75% of the rated capacity of the crane or derrick or (2) requires the use of more than one crane or derrick.

**Cross Brace:** Wood or metal diagonal bracing used to aid in structural support between joists and beams.

**Crows Foot:** A lath set by the grade setter with markings to indicate the final grade at a certain point.

**Cup:** To warp across the grain.

**Curtain Wall:** Inside walls; they do not carry loads from roof or floors above them.

**Cutting Plane Line:** A heavy broken line with arrows, letters and numbers at each end indicating the section view that is being identified.

## D

**Dado:** A rectangular groove cut into a board across the grain.

**Dampproofing:** Moisture protection; a surfacing used to coat and protect concrete and masonry from moisture penetration.

**Datum Point:** *See* benchmark; identification of the elevation above mean sea level.

**Deceleration Device:** Any mechanism—such as rope, grab, ripstitch lanyard, specially woven lanyard, tearing or deforming lanyards or automatic self-retracting lifelines/lanyards—that serves to dissipate a substantial amount of energy during a fall arrest, or otherwise limits the energy imposed on an employee during fall arrest.

**Deceleration Distance:** The additional vertical distance a falling person travels, excluding lifeline elongation and free fall distance, before stopping, from the point at which a deceleration device begins to operate.

**Dead Load:** The weight of a structure and all its fixed components.

**Deck:** The part of a roof that covers the rafters.

**Decking Hole:** A gap or void more than 2" (5.1 cm) in its least dimension and less than 12" (30.5 cm) in its greatest dimension in a floor, roof or other walking/working surface; pre-engineered holes in cellular decking (for wires, cables, etc.) are not included in this definition.

**Deformed Bar:** A steel reinforcement bar with ridges to prevent the bar from loosening during the concrete curing process.

**Derrick Floor:** An elevated floor of a building or structure that has been designated to receive hoisted pieces of steel prior to final placement.

**Diagonal Brace:** A wood or metal member placed

# GLOSSARY (cont.)

diagonally over wood or metal framing to add rigidity at corners and at 25' of unbroken wall space.

**Dimension Line:** A line on a drawing with a measurement indicating length.

**Double Connection:** An attachment method where the connection point is intended for two pieces of steel that share common bolts on either side of a central piece.

**Double Connection Seat:** A structural attachment that, during the installation of a double connection, supports the first member while the second member is connected.

**Double Plate:** Usually refers to the practice of using two pieces of dimensional lumber for support over the top section or wall section.

**Double Trimmer:** Double joists used on the sides of openings; double trimmers are placed without regard to regular joist spacings for openings in the floor for stairs or chimneys.

**Double-Cleat Ladder:** A ladder similar in construction to a single-cleat ladder, but with a center rail to allow simultaneous two-way traffic for employees ascending or descending.

**Downspouts:** Pipes connected to the gutter to conduct rainwater to the ground or sewer.

**Drain Tile:** Usually made of plastic, generally 4" in diameter, with small holes to allow water to drain into it; laid along the foundation footing to drain the seepage into a sump or storm sewer.

**Drop Siding:** Has a special groove or edge cut into it that lets each board fit into the next board and makes the boards fit together to resist moisture and weather.

**Ductwork:** A system of pipes used to pass heated air along to all parts of a house; also used to distribute cold air for summer air conditioning.

# E

**Earthwork:** Excavating and grading.

**Easement:** A portion of land on or off a property that is set aside for utility installations.

**Eave:** The lowest edge on a gable roof.

**Eaves:** The overhang of a roof projecting the walls.

**Eaves Trough:** A gutter.

**Elevation:** An exterior or interior orthographic view of a

## GLOSSARY (cont.)

structure identifying the design and the materials to be used.

**Elevation Numbers:** The vertical distance above or below sea level.

**Embankment:** The area being filled with earth.

**Equivalent:** Alternative designs, materials or methods that the employer can demonstrate that will provide an equal or greater degree of safety for employees than the method or item specified in the standard.

**Erection Bridging:** The bolted diagonal bridging that is required to be installed prior to releasing the hoisting cables from the steel joists.

**Extension Trestle Ladder:** A self-supporting portable ladder, adjustable in length, consisting of a trestle ladder base and a vertically adjustable extension section, with a suitable means for locking the ladders together.

## F

**Face:** The exposed side of a framing or masonry unit; a type of brick; also called *common*.

**Failure:** Load refusal, breakage or separation of component parts; load refusal is the point where the structural members lose their ability to carry the loads.

**Fall Restraint System:** A fall protection system that prevents the user from falling any distance; the system is composed of either a body belt or body harness, along with an anchorage, connectors and other necessary equipment; the other components typically include a lanyard and a lifeline.

**Fascia:** A flat board covering the ends of rafters on the cornice or eaves; eave troughs are usually mounted to the fascia board.

**Feathering:** Raking new asphalt to join smoothly with the existing asphalt.

**Final Interior Perimeter:** The perimeter of a large permanent open space within a building such as an atrium or courtyard; this does not include openings for stairways, elevator shafts, etc.

**Finish:** Any material used to complete an installation that provides an aesthetic or finished appearance.

**Firebrick:** A special type of brick that is not damaged by fire; used to line the firebox in a fireplace.

**Fire Stop/Draft Stop/Fire Blocking:** A framing member used to reduce the ability of a fire's spread.

**Firewall/Fire Separation Wall/Fire Division Wall:** Any wall that is installed for the purpose of preventing the spread of fire.

**Fixed Ladder:** A ladder that cannot be readily moved or carried because it is an integral part of a building or structure; a side-step fixed ladder is a fixed ladder that requires a person getting off at the top to step to the side of the ladder side rails to reach the landing; a through fixed ladder is a fixed ladder that requires a person getting off at the top to step between the side rails of the ladder to reach the landing.

**Flashing:** Metal or plastic strips or sheets used for moisture protection in conjunction with other construction materials.

**Flat:** Any roof structure up to a 3:12 slope.

**Flue:** The passage through a chimney.

**Flush:** Even with.

**Fly Ash:** Fine, powdery coal residue used with a hydraulic (water-resistant) concrete mix.

**Footing:** The bottom-most member of a foundation; supports the full load of the structure above.

**Form:** A temporary construction member used to hold permanent materials in place.

**Formwork:** The total system of support for freshly placed or partially cured concrete, including the mold or sheeting (form) that is in contact with the concrete as well as all supporting members including shores, reshores, hardware, braces and related hardware.

**Foundation:** The base on which a house or building rests.

**Frostline:** The depth to which ground freezes in the winter.

**Furring Strips:** Strips of wood attached to concrete or stone that form a nail base for wood or paneling.

## G

**Gable:** The simplest kind of roof; two large surfaces come together at a common edge forming an inverted V.

**Galvanize:** A coating of zinc primarily used on sheet metal.

**Galvanized Iron:** Sheet metal coated with zinc.

**Gambrel Roof:** A barn-shaped roof.

**Gauge:** The thickness of metal or glass sheet material.

**Girder:** A support for joists at one end; usually placed halfway between the outside walls and runs the length of the building.

**Girt:** A Z- or C-shaped member formed from sheet steel spanning between primary framing and supporting wall material.

**Glaze:** Install glass.

**Glu-Lam:** Glue-laminated beam made from milled 2x lumber bonded together to form a beam.

**Grade:** An existing or finished elevation in earthwork; a sloped portion of a roadway; sizing of gravel and sand; the structural classification of lumber.

**Grade Break:** A change in slope from one incline ratio to another.

**Grade Lath:** A piece of lath that the surveyor or grade setter has marked to indicate the correct grade to the operators.

**Grade Pins:** Steel rods driven into the ground at each surveyor's hub; a string is stretched between them at the grade indicated on the survey stakes or a constant distance above the grade.

**Grader:** A power excavating machine with a central blade that can be angled to cast soil on either side; it has an independent hoist control on each side; also called a *blade*.

**Gravel Stop:** The edge metal used at the eaves of a built-up roof to hold the gravel on the roof.

**Green:** Uncured or set concrete or masonry; freshly cut lumber.

**Grid System:** An arrangement of metal strips that support a drop ceiling.

**Ground-Fault-Circuit-Interrupter:** An electrical receptacle installed for personal safety at outdoor locations and near water.

**Guardrail System:** A barrier erected to prevent employees from falling to lower levels.

**Guinea:** A survey marker driven to grade; it may be colored with paint or crayon; used for finishing and fine trimming; also called a *hub*.

**Gusset:** A triangular or rectangular piece of wood or metal that is usually fastened to the joint of a truss to strengthen it; used primarily in making roof trusses.

**Gutter:** A metal or wooden trough set below the eaves to catch and conduct water from rain and melting snow to a downspout.

**Gypsum:** A chalk used to make wallboard; made into a paste, inserted between two layers of paper and allowed to dry; produces a plastered wall with certain fire-resisting characteristics.

## H

**Habitable Space:** The interior areas of a residence used for eating, sleeping, living and cooking; excludes bathrooms, storage rooms, utility rooms and garages.

**Handrail:** A rail used to provide employees with a handhold for support.

**Hanger:** Metal fabrication made for the purpose of placing and supporting joists and rafters.

**Hardware:** Any component used to hang, support or position another component (e.g., door and window hardware, hangers).

**Hardwood:** Wood that comes from a tree that sheds its leaves.

**Head Joint:** The end face of a brick or concrete masonry unit to which the mortar is applied.

**Headache Ball:** A weighted hook that is used to attach loads to the hoist load line of the crane.

**Header:** A framing member used to hide the ends of joists along the perimeter of a structure; also known as a *rim joist*; the horizontal structural framing member installed over wall openings to aid in the support of the structure above; also referred to as a *lintel*.

**Header Course:** A horizontal row of brick laid perpendicular to the wall face; used to tie a double wythe brick wall together.

**Hidden Line:** A dashed line identifying portions of construction that are a part of the drawing but cannot be seen (e.g., footings on foundation plans or wall cabinetry in floor plans).

**Hip Rafters:** A member that extends diagonally from the corner of the plate to the ridge.

**Hip Roof:** A structural sloped roof design with sloped perimeters from ridge to plate line.

# GLOSSARY (cont.)

**Hole:** A void or gap 2" (5.1 cm) or more in the least dimension in a floor, roof or other walking/working surface.

**Hoisting Equipment:** Commercially manufactured lifting equipment designed to lift and position a load of known weight to a location at some known elevation and horizontal distance from the equipment's center of rotation; hoisting equipment includes but is not limited to cranes, derricks, tower cranes, barge-mounted derricks or cranes, gin poles and gantry hoist systems; a come-a-long (a mechanical device, usually consisting of a chain or cable attached at each end, that is used to facilitate movement of materials through leverage) is not considered hoisting equipment.

**Hose Bibb:** A faucet used to connect a hose.

**HVAC:** Heating, ventilating and air conditioning; term given to all heating and air conditioning systems; the mechanical portion of the Construction Specification Institute (CSI) format, division 15.

**Hydraulic Cement:** A cement used in a concrete mix capable of curing under water.

## I

**Individual-Rung/Step Ladders:** Ladders without a side rail or center rail support; such ladders are made by mounting individual steps or rungs directly to the side or wall of the structure.

**Insulation:** Any material capable of resisting thermal, sound or electrical transmission.

**Insulation Resistance:** The R factor in insulation calculations.

## J

**Jack Rafter:** A part of the roof structure raftering that does not extend the full length from the ridge beam to the top plate.

**Jacking Operation:** Lifting vertically a slab (or group of slabs) from one location to another—for example, from the casting location to a temporary (parked) location, or from a temporary location to another temporary location, or to the final location in the structure—during a lift-slab construction operation.

**Jamb:** The part that surrounds a door window frame; usually made of two vertical pieces and a horizontal piece over the top.

**Job-Made Ladder:** A ladder that is fabricated by employees, typically at the construction site, and is not commercially manufactured; this definition does not apply to any individual-rung/step ladders.

**Joint Compound:** Material used with a paper of fiber tape for sealing indentations and breaks in drywall construction.

**Joist:** A structural, horizontal framing member used for floor and ceiling support systems.

**Joist Hangers:** Metal brackets that hold up the joist; they are nailed to the girder, and the joist fits into the bracket.

### K

**Key:** A depression made in a footing so that the foundation or wall can be poured into the footing, preventing the wall or foundation from moving during changes in temperature or settling of the building.

**Kicker Blocks:** Cement poured behind each bend or angle of water pipe for support; also called *thrust blocks.*

**Kiln-Dried Lumber:** Lumber that is seasoned under controlled conditions, removing 6% to 12% of the moisture in green lumber.

**King Stud:** A full-length stud from the bottom plate to the top plate supporting both sides of a wall opening.

**Knee Wall:** Vertical framing members supporting and shortening the span of the roof rafters.

### L

**Ladder Stand:** A mobile fixed-size self-supporting ladder consisting of a wide flat tread ladder in the form of stairs; the assembly may include handrails.

**Lanyard:** A flexible line of rope, wire rope or strap that generally has a connector at each end for connecting the body belt or body harness to a deceleration device, lifeline or anchorage.

**Lateral:** Underground electrical service.

**Lath:** Backup support for plaster; may be of wood, metal or gypsum board.

**Lavatory:** Bathroom; vanity basin.

**Lay-In Ceiling:** A suspended ceiling system.

**Leach Line:** A perforated pipe used as a part of a septic system to allow liquid overflow to dissipate into the soil.

**Leading Edge:** The unprotected side and edge of a floor, roof or formwork for a floor or other walking/working surface (such as deck) that changes location as additional floor, roof, decking or formwork sections are placed, formed or constructed.

**Ledger:** A structural framing member used to support ceiling and roof joists at the perimeter walls.

**Level-Transit:** An optical device that is a combination of a level and a means for checking vertical and horizontal angles.

**Lifeline:** A component consisting of a flexible line for connection to an anchorage at one end to hang vertically (vertical lifeline), or for connection to anchorages at both ends to stretch horizontally (horizontal lifeline), and that serves as a means for connecting other components of a personal fall arrest system to the anchorage.

**Lift Slab:** A method of concrete construction in which floor and roof slabs are cast on or at ground level and, using jacks, are lifted into position.

**Limited Access Zone:** An area alongside a masonry wall that is under construction and clearly demarcated to limit access by employees.

**Lift:** Any layer of material or soil placed on another.

**Live Load:** Any movable equipment or personal weight to which a structure is subjected.

**Load:** The weight of a building.

**Load Conditions:** The conditions under which a roof must perform.

**Lockset:** The doorknob and associated locking parts inserted in a door.

**Lower Levels:** Those areas to which an employee can fall from a stairway or ladder; such areas include ground levels, floors, roofs, ramps, runways, excavations, pits, tanks, material, water, equipment and similar surfaces; it does not include the surface from which the employee falls.

**Low-Slope Roof:** A roof having a slope less than or equal to 4 in 12 (vertical to horizontal).

# M

**Masonry:** Manufactured materials of clay, concrete and stone.

**Mastic:** An adhesive used to hold tiles in place; also refers to adhesives used to glue many types of materials in the building process.

**Mat:** Asphalt as it comes out of a spreader box or paving machine in a smooth, flat form.

**Maximum Density and Optimum Moisture:** The highest point on the moisture density curve; considered the best compaction of the soil.

**Maximum Intended Load:** The total load of all employees, equipment, tools, materials, transmitted loads and other loads anticipated to be applied to a ladder component at any one time.

**MEE Pipe:** Pipe that has been milled on each end and left rough in the center; MEE stands for "milled each end."

**Membrane Roofing:** Built-up roofing.

**Mesh:** Common term for welded wire fabric, plaster lath.

**Metal Decking:** A commercially manufactured, structural grade, cold rolled metal panel formed into a series of parallel ribs; this subpart includes metal floor and roof decks, standing seam metal roofs, other metal roof systems and other products such as bar gratings, checker plate and expanded metal panels; after installation and proper fastening, these decking materials serve a combination of functions, including an element designed in combination with the structure to resist, distribute and transfer loads and stiffen the structure and provide a diaphragm action, walking/working surface, form for concrete slabs, support for roofing systems and finished floor or roof.

**Mil:** 0.001"

**Military Specifications:** Details that the military writes for the products it buys from the manufacturers.

**Minute:** 1/60th of a degree.

**MOA Pipe:** Pipe that has been milled end to end; MOA stands for "milled over all" and allows easier joining of the pipe if the length must be cut to fit.

**Moisture Barrier:** A material used for the purpose of resisting exterior moisture penetration.

**Moisture Density Curve:** A graph plotted from tests to determine at what point of added moisture the maximum density will occur.

**Moldings:** Trim mounted around windows, floors and doors as well as closets.

**Monolithic Concrete:** Concrete placed as a single unit, including turndown footings.

**Mortar:** A concrete mix especially used for bonding masonry units.

**Multiple Lift Rigging:** A rigging assembly manufactured by wire rope rigging suppliers that facilitates the attachment of up to five independent loads to the hoist rigging of a crane.

## N

**Natural Grade:** Existing or original grade elevation of a property.

**Natural Ground:** The original ground elevation before any excavation has been done.

**Nominal Size:** The original cut size of a piece of lumber prior to milling and drying; size of masonry unit, including mortar bed and head joint.

**Non-Bearing:** Not supporting any structural load.

**Nosing:** That portion of a tread projecting beyond the face of the riser immediately below.

**Nuclear Test:** A test to determine soil compaction by sending nuclear impulses into the compacted soil and measuring the returned impulses reflected from the compacted particles.

## O

**On Center:** The distance between the centers of two adjacent components.

**Open Web Joist:** A roof joist made of wood or steel construction with a top chord and bottom chord connected by diagonal braces bolted or welded together.

**Opening:** A gap or void 12" (30.5 cm) or more in its least dimension in a floor, roof or other walking/working surface; for the purposes of this subpart, skylights and smoke domes that do not meet the strength requirements of § 1926.754(e)(3) shall be regarded as openings.

**Opening:** A gap or void 30" (76 cm) or higher and 18" (46 cm) or wider, in a wall or partition, through which employees can fall to a lower level.

## P

**Package Air Conditioner or Boiler:** An air conditioner or boiler in which all components are packaged into a single unit.

**Pad:** The base materials upon which to place the concrete footing and/or slab.

**Parapet:** An extension of an exterior wall above the line of the roof.

**Parging:** A thin moisture protection coating of plaster or mortar over a masonry wall.

**Partition:** An interior wall separating two rooms or areas of building; usually non-bearing.

**Penny:** The unit of measure of the nails used by carpenters.

**Perimeter:** The outside edges of a plot of land or building; it represents the sum of all the individual sides.

**Perimeter Insulation:** Insulation placed around the outside edges of a slab.

**Permanent Floor:** A structurally completed floor at any level or elevation (including slab on grade).

**Personal Fall Arrest System:** A system used to arrest an employee in a fall from a working level; a personal fall arrest system consists of an anchorage, connectors and a body harness, and it may include a lanyard, deceleration device, lifeline or suitable combination of these; the use of a body belt for fall arrest is prohibited.

**Personal Fall Arrest System:** A system including but not limited to an anchorage, connectors and a body belt or body harness used to arrest an employee in a fall from a working level; as of January 1, 1998, the use of a body belt for fall arrest is prohibited.

**Pile:** A steel or wooden pole driven into the ground sufficiently to support the weight of a wall and building.

**Pillar:** A pole or reinforced wall section used to support the floor and consequently the building.

**Pitch:** The slant or slope from the ridge to the plate.

**Plan View:** A bird's-eye view of a construction layout cut at 5' above finish floor level.

**Plaster:** A mixture of cement, water and sand.

## GLOSSARY (cont.)

**Plate:** A roof member that has the rafters fastened to it at their lower ends.

**Platform Framing:** Also known as *western framing*; structural construction in which all studs are only one story high with joists over.

**Point of Access:** All areas used by employees for work-related passage from one area or level to another; such open areas include doorways, passageways, stairway openings, studded walls and various other permanent or temporary openings used for such travel.

**Point Of Beginning:** The point on a property from which all measurements and azimuths are established.

**Polyvinyl Chloride:** A plastic material commonly used for pipe and plumbing fixtures.

**Portable Ladder:** A ladder that can be readily moved or carried.

**Portland Cement:** One variety of cement produced from burning various materials such as clay, shale and limestone, producing a fine gray powder; the basis of concrete and mortar.

**Positioning Device System:** A body belt or body harness system rigged to allow an employee to be supported on an elevated vertical surface, such as a wall, and work with both hands free while leaning backward.

**Post:** A structural member with a longitudinal axis that is essentially vertical and (1) weighs 300 pounds or less and is axially loaded (a load presses down on the top end), or (2) is not axially loaded, but is laterally restrained by the above member; posts typically support stair landings, wall framing, mezzanines and other substructures.

**Post-And-Beam Construction:** A type of wood frame construction that uses timber for the structural support.

**Post-Tensioning:** The application of stretching steel cables embedded in a concrete slab to aid in strengthening the concrete.

**Precast Concrete:** Concrete members (such as walls, panels, slabs, columns and beams) that have been formed, cast and cured prior to final placement in a structure.

# GLOSSARY (cont.)

**Prehung:** Refers to doors or windows that are already mounted in a frame and are ready for installation as a complete unit.

**Pressure Treatment:** Impregnating lumber with a preservative chemical under pressure in a tank.

**Primer:** The first coat of paint or glue when more than one coat will be applied.

**Project Structural Engineer Of Record:** The registered, licensed professional responsible for the design of structural steel framing and whose seal appears on the structural contract documents.

**Purlin:** A Z- or C-shaped member formed from sheet steel spanning between primary framing and supporting roof material.

**Purlin:** A horizontal framing member spanning between rafters.

## Q

**Qualified Person:** One who, by possession of a recognized degree, certificate or professional standing, or who by extensive knowledge, training and experience, has successfully demonstrated the ability to solve or resolve problems relating to the subject matter, work or project.

**Quarry Tile:** An unglazed clay or shale flooring material produced by the extrusion process.

**Quick Set:** A fast-curing cement plaster.

## R

**R Factor:** The numerical rating given any material that is able to resist heat transfer for a specific period of time.

**R Values:** The unit that measures the effectiveness of insulation; the higher the number, the better the insulation qualities of the materials.

**Rabbet:** A groove cut in or near the edge of a piece of lumber to fit the edge of another piece.

**Raceway:** Any partially or totally enclosed container for placing electrical wires (conduit, and so on).

**Rafter:** The framing member extending from the ridge or hip to the top plate in a sloped roof.

**Rebar:** A reinforcement steel rod in a concrete footing.

**Reshoring:** The construction operation in which shoring

# GLOSSARY (cont.)

equipment (also called reshores or reshoring equipment) is placed as the original forms and shores are removed in order to support partially cured concrete and construction loads.

**Resilient Flooring:** Flooring made of plastic rather than wood products.

**Ridge:** The highest point on a sloped roof.

**Ridge Board:** A horizontal member that connects the upper ends of the rafters on one side to the rafters on the opposite side.

**Right-Of-Way Line:** A band on the side of a road marking the limit of the construction area and usually, the beginning of private property.

**Rise:** In roofing, rise is the vertical distance between the top of the double plate and the center of the ridge board; in stairs, it is the vertical distance from the top of a stair tread to the top of the next tread.

**Riser Height:** The vertical distance from the top of a tread to the top of the next higher tread or platform/landing or the distance from the top of a platform/landing to the top of the next higher tread or platform/landing.

**Riser:** The vertical part at the edge of a stair.

**Roll Roofing:** A type of built-up roofing material made of a mixture of rag, paper and asphalt.

**Rope Grab:** A deceleration device that travels on a lifeline and automatically, by friction, engages the lifeline and locks to arrest a fall.

**Roof Pitch:** The ratio of total span to total rise expressed as a fraction.

**Rough Opening:** A large opening made in a wall frame or roof frame to allow the insertion of a door or window.

**RS:** The reference stake, from which measurements and grades are established.

**Run:** The shortest horizontal distance measured from a plumb line through the center of the ridge to the outer edge of the plate of a roof.

## S

**Safety Deck Attachment:** An initial attachment that is used to secure an initially placed sheet of decking to keep proper alignment and bearing with structural support members.

**Safety-Monitoring System:** A safety system in which a competent person is responsible for recognizing and warning employees of fall hazards.

**Sand Cone Test:** Determines the compaction level of soil by removing an unknown quantity of soil and replacing it with a known quantity of sand.

**Scabs:** Boards used to join the ends of a girder.

**Schematic:** A one-line drawing for electrical circuitry or isometric plumbing diagrams.

**Scissors Truss:** A truss constructed to the roof slope at the top chord with the bottom chord designed with a lower slope for interior vaulted or cathedral ceilings.

**Scraper:** A digging, hauling and grading machine that has a cutting edge, carrying bowl, movable front wall and dumping mechanism.

**Scratch Coat:** The first coat of plaster placed over the lath in a three-coat plaster system.

**Scupper:** An opening in a parapet wall attached to a downspout for water drainage from the roof.

**Scuttle:** Attic or roof access with cover or door.

**Sealant:** A material used to seal off openings against moisture and air penetration.

**Section:** A vertical drawing showing architectural or structural interior design developed at the point of a cutting-plane line on a plan view; the section may be transverse—the gable end—or longitudinal—parallel to the ridge.

**Self-Retracting Lifeline/Lanyard:** A deceleration device containing a drum-wound line that can be slowly extracted from, or retracted onto, the drum under minimal tension during normal employee movement and that, after onset of a fall, automatically locks the drum and arrests the fall.

**Seismic Design:** Construction designed to withstand earthquakes.

**Septic System:** A waste system used in lieu of a sewer system that includes a line from the structure to a tank and a leach field.

**Shakes:** Shingles made of handsplit wood; in most cases western cedar.

**Shear Connector:** Headed steel studs, steel bars, steel lugs and similar devices that are attached to a structural member for the purpose of achieving composite action with concrete.

**Shear Wall:** A wall construction designed to withstand shear pressure caused by wind or earthquake.

**Sheathing:** The outside layer of wood applied to studs to close up a house or wall; also used to cover the rafters and make a base for the roofing.

**Sheepsfoot Roller:** A compacting roller with feet expanded at their outer tips; used in compacting soil.

**Shore:** A supporting member that resists a compressive force imposed by a load.

**Shoring:** Temporary support made of metal or wood used to support other components.

**Side-Step Fixed Ladder:** *See* fixed ladder.

**Sill:** A piece of wood that is anchored to the foundation.

**Single-Cleat Ladder:** A ladder consisting of a pair of side rails, connected together by cleats, rungs or steps.

**Single-Rail Ladder:** A portable ladder with rungs, cleats or steps mounted on a single rail instead of the normal two rails used on most other ladders.

**Sinker Nail:** Used for laying subflooring; the head is sloped toward the shank but is flat on top.

**Size:** A special coating used for walls before wallpaper is applied; it seals the walls and allows the wallpaper paste to attach itself to the wall and paper without adding undue moisture to the wall.

**Slab-On-Grade:** The foundation construction for a structure with no crawl space or basement.

**Slump:** The consistency of concrete at the time of placement.

**Snaphook:** A connector consisting of a hook-shaped member with a normally closed keeper, or similar arrangement, that may be opened to permit the hook to receive an object and, when released, automatically closes to retain the object.

**Soffit:** A covering for the underside of the overhang of a roof.

**Soleplate:** A $2 \times 4$ or $2 \times 6$ used to support studs in a horizontal position; it is placed against the flooring and nailed into position onto the subflooring.

**Span:** The horizontal distance between exterior bearing walls in a transverse section.

## GLOSSARY (cont.)

**Specifications:** The written instructions detailing the requirements of construction for a project.

**Spiral Stairway:** A series of steps attached to a vertical pole and progressing upward in a winding fashion within a cylindrical space.

**Spoil Site:** The area used to dispose of unsuitable or excess excavation material.

**Spreader:** Braces used across the top of concrete forms.

**Square:** A roof-covering area; a square consists of 100 square feet of area.

**Stain:** A paint-like material that imparts a color to wood.

**Stair Rail System:** A vertical barrier erected along the unprotected sides and edges of a stairway to prevent employees from falling to lower levels; the top surface of a stairrail system may also be a handrail.

**Steep Roof:** A roof having a slope greater than 4 in 12 (vertical to horizontal).

**Steel Erection:** The construction, alteration or repair of steel buildings, bridges and other structures, including the installation of metal decking and all planking used during the process of erection.

**Steel Joist:** An open web, secondary load-carrying member of 144' (43.9 m) or less designed by the manufacturer and used for the support of floors and roofs; this does not include structural steel trusses or cold-formed joists.

**Steel Joist Girder:** An open web, primary load-carrying member designed by the manufacturer and used for the support of floors and roofs; this does not include structural steel trusses.

**Steel Truss:** An open web member designed of structural steel components by the project structural engineer of record; for the purposes of this subpart, a steel truss is considered equivalent to a solid web structural member.

**Step Stool:** A self-supporting, foldable, portable ladder, nonadjustable in length, 32" or less in overall size, with flat steps and without a pail shelf, designed to be climbed on the ladder top cap as well as all steps; the side rails may continue above the top cap.

**Stepped Footing:** A footing that may be located on a number of levels.

**Stool:** The flat shelf that rims the bottom of a window frame on the inside of a wall.

**Stress Skin Panels:** Large prebuilt panels used as walls, floors and roof decks; built in a factory and hauled to the building site.

**String Line:** A nylon line usually strung tightly between supports to indicate both direction and elevation; used in checking grades or deviations in slopes or rises.

**Strip Flooring:** Wooden strips that are applied perpendicular to the joists.

**Strongbacks:** Braces used across ceiling joints that help align, space and strengthen joists for drywall installation.

**Structural Steel:** Heavy steel members larger than 12-gauge identified by their shapes.

**Structural Steel:** A steel member or a member made of a substitute material (such as, fiberglass, aluminum or composite members); these members include steel joists, joist girders, purlins, columns, beams, trusses, splices, seats, metal decking, girts, all bridging and cold formed metal framing that is integrated with the structural steel framing of a building.

**Stucco:** A type of finish used on the outside of a building; a masonry finish that can be put on over any type of wall; applied over a wire mesh nailed to the wall.

**Stud:** The vertical boards (usually $2 \times 4$ or $2 \times 6$) that make up the walls of a building.

**Subfloor:** A platform that supports the rest of the structure; also referred to as underlayment.

**Subgrade:** The uppermost level of material placed in embankment or left at cuts in the normal grading of a road bed.

**Summit:** The highest point of any area or grade.

**Super:** A continuous slope in one direction on a road.

**Swale:** A shallow dip made to allow the passage of water.

**Sway Brace:** A piece of $2 \times 4$ or similar material used to temporarily brace a wall from wind until it is secured.

**Swedes:** A method of setting grades at a center point by sighting across the tops of three lath; two lath are placed at a known correct elevation, and the third is adjusted until it is at the correct elevation.

**Symbol:** A pictorial representation of a material or component on a plan.

**Systems-Engineered Metal Building:** A metal, field-assembled building system consisting of framing, roof and wall coverings; typically, many of these components are cold-formed shapes; these individual parts are fabricated in one or more manufacturing facilities and shipped to the job site for assembly into the final structure; the engineering design of the system is normally the responsibility of the systems-engineered metal building manufacturer.

# T

**Tail Joist:** A short beam or joist supported in a wall on one end and by a header on the other.

**Tail/Rafter Tail:** The portion of a roof rafter extending beyond the plate line.

**Tamp:** Pack tightly; usually refers to making sand tightly packed or making concrete mixed properly in a form to get rid of air pockets that may form with a quick pouring.

**Tangent:** A straight line from one point to another that passes over the edge of a curve.

**Tank:** A container for holding gases, liquids or solids.

**Taping And Bedding:** Drywall finishing; taping is the application of a strip of specially prepared tape to drywall joints; bedding means embedding the tape in the joint to increase its structural strength.

**Temporary Service Stairway:** A stairway where permanent treads and/or landings are to be filled in at a later date.

**Tensile Strength:** The maximum stretching of a piece of metal (rebar and so on) before breaking; calculated in kps.

**Tensioning:** Pulling or stretching of steel tendons to aid in reinforcement of concrete.

**Terrazzo:** A mixture of concrete, crushed stone, calcium shells and/or glass, polished to form a tile-like finish.

**Texture Paint:** A thick paint that will leave a texture or pattern; can be shaped to cover cracked ceilings or walls or beautify an otherwise dull room.

**Thermal Ceilings:** Ceilings that are insulated with batts of insulation to prevent loss of heat or cooling.

**Through Fixed Ladder:** A fixed ladder that requires a person getting off at the top to step between the side rails of the ladder to reach the landing.

**Tie:** A soft metal wire that is twisted around a rebar or reinforcement rod and chair to hold the rod in place till concrete is poured.

**Tied Out:** The process of determining the fixed location of existing objects (manholes, meter boxes, etc.) in a street so that they may be uncovered and raised after paving.

**Toeboard:** A low protective barrier that prevents material and equipment from falling to lower levels and that protects personnel from falling.

**Toe Of Slope:** The bottom of an incline.

**Top Chord:** The top-most member of a truss.

**Top Plate:** The horizontal framing member fastened to the top of the wall studs; usually doubled.

**Track Loader:** A loader on tracks used for filling and loading materials.

**Tread Depth:** The horizontal distance from front to back of a tread (excluding nosing, if any).

**Tread:** The part of a stair on which people step.

**Tremie:** A pipe through which concrete may be deposited under water.

**Trimmer:** A piece of lumber, usually a 2 × 4, that is shorter than the stud or rafter but is used to fill in where the longer piece would have been normally spaced except for the window or door opening or some other opening in the roof or floor or wall.

**Truss:** A prefabricated, sloped roof system incorporating a top chord, bottom chord and bracing.

## U

**Underlayment:** Also referred to as the subfloor; used to support the rest of the building; may also refer to the sheathing used to cover rafters and serve as a base for roofing.

**Unfaced Insulation:** Padding that does not have a facing or plastic membrane over one side of it; it has to be placed on top of existing insulation; if used in a wall, it has to be covered by a plastic film to ensure a vapor barrier.

**Unprotected Sides And Edges:** Any side or edge (except at entrances to points of access) of a stairway where there is no stair rail system or wall 36" (0.9 m) or more in height, and any side or edge (except at entrances to points of access) of a stairway landing, or a ladder platform where there is no wall or guardrail system 39" (1 m) or more in height.

## V

**Valley:** The area of a roof where two sections come together and form a depression.

**Valley Rafters:** A rafter that extends diagonally from the plate to the ridge at the line of intersection of two roof surfaces.

**Vapor Barrier:** The same as a moisture barrier.

**Veneer:** A thin layer or sheet of wood.

**Vent:** Usually a hole in the eaves or soffit to allow the circulation of air over an insulated ceiling; usually covered with a piece of metal or screen.

**Vent Stack:** A system of pipes used for air circulation and to prevent water from being suctioned from the traps in the waste disposal system.

**Ventilation:** The exchange of air or the movement of air through a building; may be done naturally through doors and windows or mechanically by motor-driven fans.

**Vibratory Roller:** A self-powered or towed compacting device that mechanically vibrates while it rolls.

**Vertical Slip Forms:** Forms that are jacked vertically during the placement of concrete.

## W

**Waler:** A 2× piece of lumber installed horizontally to form-work to give added stability and strength to the forms.

**Walking/Working Surface:** Any surface, horizontal or vertical, on which an employee walks or works, including but not limited to floors, roofs, ramps, bridges, runways, formwork and concrete reinforcing steel; does not include ladders, vehicles or trailers on which employees must be located to perform their work duties.

## GLOSSARY (cont.)

**Warning Line System:** A barrier erected on a roof to warn employees that they are approaching an unprotected roof side or edge and that designates an area in which roofing work may take place without the use of guardrail, body belt or safety net systems to protect employees in the area.

**Water-Cement Ratio:** The ratio between the weight of water to cement.

**Waterproofing:** Preferably called *moisture protection*; materials used to protect below- and on-grade construction from moisture penetration.

**Water Table:** The amount of water that is present in any area; the moisture may be from rain or snow.

**Welded Wire Fabric:** A reinforcement used for horizontal concrete strengthening.

**Wind Lift (Wind Load):** The force exerted by the wind against a structure caused by the movement of the air.

**Winder:** Fan-shaped steps that allow the stairway to change direction without a landing.

**Window Apron:** The flat part of the interior trim of a window; located next to the wall and directly beneath the window stool.

**Window Stool:** The flat narrow shelf that forms the top member of the interior trim at the bottom of a window.

**Windrow:** The spill-off from the ends of a dozer or grader blade that forms a ridge of loose material; a windrow may be deliberately placed for spreading by another machine.

**Wythe:** A continuous masonry wall width.

## XYZ

**X Brace:** A cross brace for joist construction.

## Z

**Zinc:** Non-corrosive metal used for galvanizing other metals.

## ABBREVIATIONS

| | | | |
|---|---|---|---|
| **A** | area | **CC** | center to center, cubic centimeter |
| **AB** | anchor bolt | | |
| **AC** | alternate current | **CEM** | cement |
| **A/C** | air conditioning | **CER** | ceramic |
| **ACT** | acoustical ceiling tile | **CFM** | cubic feet per minute |
| **AFF** | above finish floor | **CIP** | cast-in-place, concrete-in-place |
| **AGGR** | aggregate | | |
| **AIA** | American Institute of Architects | **CJ** | ceiling joist, control joint |
| | | **CKT** | circuit (electrical) |
| **AL, ALUM** | aluminum | **CLG** | ceiling |
| **AMP** | ampere | **CMU** | concrete masonry unit |
| **APPROX** | approximate | | |
| **ASPH** | asphalt | **CO** | cleanout (plumbing) |
| **ASTM** | American Society for Testing Materials | **COL** | column |
| | | **CONC** | concrete |
| | | **CONST** | construction |
| **AWG** | American wire gauge | **CONTR** | contractor |
| | | **CU FT (ft³)** | cubic foot (feet) |
| | | **CU IN (in³)** | cubic inch(es) |
| **BD** | board | **CU YD (yd³)** | cubic yard(s) |
| **BD FT (BF)** | board foot (feet) | | |
| | | **d** | pennyweight (nail) |
| **BLDG** | building | | |
| **BLK** | black, block | **DC** | direct current (elec.) |
| **BLKG** | blocking | | |
| **BM** | board measure | **DET** | detail |

| | | | |
|---|---|---|---|
| DIA | diameter | FOS | face of stud, flush on slab |
| DIAG | diagonal | | |
| DIM | dimension | FT | foot, feet |
| DN | down | FTG | footing |
| DO | ditto (same as) | FURN | furnishing, furnace |
| DS | downspout | | |
| DWG | drawing | FX GL (FX) | fixed glass |
| | | | |
| E | East | GA | gauge |
| EA | each | GAL | gallon |
| ELEC | electric, electrical | GALV | galvanize(d) |
| ELEV | elevation, elevator | GD | ground (earth/electric) |
| ENCL | enclosure | GI | galvanized iron |
| EXCAV | excavate, excavation | GL | glass |
| | | GL BLK | glass block |
| EXT | exterior | GLB, GLU-LAM | glue-laminated beam |
| FDN | foundation | GRD | grade, ground |
| FIN | finish | GWB | gypsum wall board |
| FIN FLR | finish floor | | |
| FIN GRD | finish grade | GYP | gypsum |
| FL, FLR | floor | | |
| FLG | flooring | HB | hose bibb |
| FLUOR | fluorescent | HDR | header |
| FOB | free-on-board, factory-on-board | HDW | hardware |
| | | HGT/HT | height |
| FOM | face of masonry | HM | hollow metal |

## ABBREVIATIONS *(cont.)*

| | | | |
|---|---|---|---|
| **HORIZ** | horizontal | **MAT'L** | material |
| **HP** | horsepower | **MAX** | maximum |
| **HWH** | hot water heater | **MBF/MBM** | thousand board feet, thousand board measure |
| **ID** | inside diameter | | |
| **IN** | inch(es) | **MECH** | mechanical |
| **INSUL** | insulation | **MISC** | miscellaneous |
| **INT** | interior | **MK** | mark (identifier) |
| **J, JST** | joist | **MO** | momentary (electrical contact), masonry opening |
| **JT** | joint | | |
| **KG** | kilogram | **N** | North |
| **KL** | kiloliter | **NEC** | National Electric Code |
| **KM** | kilometer | | |
| **KWH** | kilowatt-hour | **NIC** | not in contract |
| | | **NOM** | nominal |
| **L** | left, line | | |
| **LAU** | laundry | **O/A** | overall (measure) |
| **LAV** | lavatory | **OC (O/C)** | on center |
| **LBR** | labor | **OD** | outside diameter |
| **LDG** | landing, leading | **OH** | overhead |
| **LDR** | leader | **O/H** | overhang (eave line) |
| **LEV/LVL** | level | | |
| **LIN FT (LF)** | lineal foot (feet) | **OPG** | opening |
| **LGTH** | length | **OPP** | opposite |
| **LH** | left hand | | |
| **LITE/LT** | light (window pane) | **PC** | piece |
| | | **PLAS** | plastic |

| | | | |
|---|---|---|---|
| **PLAST** | plaster | **SERV** | service (utility) |
| **PLT** | plate (framing) | **SEW** | sewer |
| **PR** | pair | **SHTHG** | sheathing |
| **PREFAB** | prefabricate(d)(tion) | **SIM** | similar |
| **PTN** | partition | **SP** | soil pipe (plumbing) |
| **PVC** | polyvinylchloride pipe | **SPEC** | specification |
| | | **SQ FT (ft²)** | square foot (feet) |
| **QT** | quart | **SQ IN (in²)** | inch(es) |
| **QTY** | quantity | **SQ YD (yd²)** | square yard(s) |
| | | **STA** | station |
| **R** | right | **STD** | standard |
| **RD** | road, round, roof drain | **STIR** | stirrup (rebar) |
| **REBAR** | reinforced steel bar | **STL** | steel |
| | | **STR/ST** | street |
| **RECEPT** | receptacle | **STRUCT** | structural |
| **REINF** | reinforce(ment) | **SUSP CLG** | suspended ceiling |
| **REQ'D** | required | | |
| **RET** | retain(ing), return | **SYM** | symbol, symmetric |
| **RF** | roof | | |
| **RFG** | roofing (materials) | **SYS** | system |
| **RH** | right hand | | |
| | | **T&G** | tongue and groove |
| **S** | South | **THK** | thick |
| **SCH/SCHED** | schedule | **TOB** | top of beam |
| **SECT** | section | **TOC** | top of curb |

## ABBREVIATIONS (cont.)

| | | | |
|------|----------------|--------|------------------------|
| **TOF** | top of footing | **W** | West |
| **TOL** | top of ledger | **w/** | with |
| **TOP** | top of parapet | **w/o** | without |
| **TOS** | top of steel | **WC** | water closet (toilet) |
| **TR** | tread, transition | | |
| **TRK** | track, truck | **WDW** | window |
| **TYP** | typical | **WI** | wrought iron |
| | | **WP** | waterproof, weatherproof |
| **UF** | underground feeder (electrical) | **WT/WGT** | weight |
| | | **YD** | yard |
| **USE** | underground service entrance cable (electrical) | **Z** | zinc |

## ABOUT THE AUTHOR

Paul Rosenberg has an extensive background in the construction, data, electrical, HVAC and plumbing trades. He is a leading voice in the electrical industry with years of experience from an apprentice to a project manager. Paul has written for all of the leading electrical and low voltage industry magazines and has authored more than 30 books.

In addition, he wrote the first standard for the installation of optical cables (ANSI-NEIS-301) and was awarded a patent for a power transmission module. Paul currently serves as contributing editor for *Power Outlet Magazine,* teaches for Iowa State University and works as a consultant and expert witness in legal cases. He speaks occasionally at industry events.